Colliery landscapes

An aerial survey of the deep-mined coal industry in England

Colliery landscapes

An aerial survey of the deep-mined coal industry in England

Shane Gould and Ian Ayris

ENGLISH HERITAGE

1995

First published 1995 by English Heritage, 23 Savile Row, London W1X 1AB

A catalogue record is available from the British Library

Printed by Billington Press Limited

ISBN 1 85074 508 0

Edited by Kate Macdonald
Designed by Nick Cannan

Contents

List of figures......6

Preface......7

Introduction......9

Editorial note......11

Acknowledgements......11

The coal industry: a brief historical survey of mining technology......12
by Shane Gould

Gazetteer of coalfields *by Ian M Ayris*......17

Western Area coalfields 1-13......22

North East Area coalfields 14 - 3......38

Yorkshire coalfields
North Yorkshire Area 32-39......60
South Yorkshire Area 40-49......72
Barnsley Area 50-59......83
Doncaster Area 60-68......95

Midlands coalfields
North Derbyshire Area 69-75......104
North Nottinghamshire Area 76-84......112
South Nottinghamshire Area 85-90......121
South Midlands Area 91-93......130

Glossary of mining terms......134

Alphabetical index of sites included in the aerial survey......137

List of figures

Except where otherwise indicated, all photographs are by RCHME, Crown Copyright 1995.

1 Shireoaks Colliery, South Yorkshire, 1961.......8
2 Shireoaks Colliery, South Yorkshire, 1992.......8
3 Parkside Colliery, Newton-le-Willows...........10
4 Hem Heath Colliery, Stoke-on-Trent, 1951...13
5 Hem Heath Colliery, Stoke-on-Trent, 1992...13
6 Ireland Colliery, North Derbyshire................16
7 Frickley/South Elmshall combined mine, Doncaster Area...17
8 Map 1: NCB Areas.......................................21
9 Bolsover Colliery, North Derbyshire Area, 1947 (English Heritage)...21
10 Map 2: Western Area.................................22
11 Bickershaw Colliery, Western Area, 1970s (English Heritage).......................................22
12 1 Agecroft Colliery, Western Area...............23
13 2.1 Bickershaw Colliery, Western Area.........24
14 2.2 Bickershaw Colliery, Western Area.........25
15 3 Bold Colliery, Western Area..................... 26
16 4 Cronton Colliery, Western Area................27
17 5 Florence Colliery, Western Area...............28
18 6 Golborne Colliery, Western Area.............29
19 7 Haig Colliery, Western Area.....................30
20 8 Hem Heath Colliery, Western Area...........31
21 9 Lea Hall Colliery, Western Area...............32
22 10 Parkside Colliery, Western Area..............33
23 11 Parsonage Colliery, Western Area...........34
24 12.1 Silverdale Colliery, Western Area.........35
25 12.2 Silverdale Colliery, Western Area.........36
26 13 West Cannock No 5 Colliery, Western Area... 37
27 Map 3: North East Area.............................38
28 Eppleton Colliery, North East Area, 1970s (English Heritage).......................................38
29 14 Ashington Colliery, North East Area.......39
30 15 Bates Colliery, North East Area.............40
31 16 Brenkley Colliery, North East Area........ 41
32 17.1 Dawdon Colliery, North East Area......42
33 17.2 Dawdon Colliery, North East Area......43
34 18.1 Easington Colliery, North East Area....44
35 18.2 Easington Colliery, North East Area....45
36 19 Elemore Colliery, North East Area.........46
37 20 Ellington Colliery, North East Area........47
38 21 Epleton Colliery, North East Area..........48
39 22 Houghton Colliery, North East Area.......49
40 23 Hylton and Wearmouth Collieries, North East Area..50
41 24 Lynemouth Colliery, North East Area.....51
42 25 Murton Colliery, North East Area..........52
43 26.1 Seaham Colliery, North East Area.......53
44 26.2 Seaham Colliery, North East Area.......54
45 27 Shilbottle Colliery, North East Area.......55
46 28 South Hutton Colliery/Hawthorn Combined Mine, North East Area...............56
47 29 Vane Tempest Colliery, North East Area..57
48 30 Westoe Colliery, North East Area...........58
49 31 Woodhorn Colliery, North East Area......59
50 Map 4: Yorkshire Coalfields Area................ 60
51 Orgreave Colliery, South Yorkshire Area, 1977 (English Heritage).......................................60
52 32 Allerton Bywater Colliery, North Yorkshire Area..61
53 33 Fryston Colliery, North Yorkshire Area..62
54 34.1 Kellingley Colliery, North Yorkshire Area..63
55 34.2 Kellingley Colliery, North Yorkshire Area..64
56 34.3 Kellingley Colliery, North Yorkshire Area..65
57 35 Ledston Luck Colliery, North Yorkshire Area..66
58 36 Prince of Wales Colliery, North Yorkshire Area..67
59 37 Selby Project/Gascoigne Wood Drift, North Yorkshire Area...68
60 38.1 Selby Project/Shaft Mines, North Yorkshire Area... 69
61 38.2 Selby Project/Riccall Mine...................70
62 38.3 Selby Project/Stillingfleet Mine...........70
63 38.4 Selby Project/Whitemoor Mine...........71
64 38.5 Selby Project/Wistow Mine.................71
65 39 Sharlston Coliery, North Yorkshire Area..72
66 40 Cortonwood Colliery, South Yorkshire Area.. 73
67 41 Dinnington Colliery, South Yorkshire Area..74
68 42 Kiveton Park Colliery, South Yorkshire Area.. 75
69 43 Maltby Colliery, South Yorkshire Area....76
70 44 Manton Colliery, South Yorkshire Area.. 77
71 45 Orgreave Colliery, South Yorkshire Area..78
72 46 Silverwood Colliery, South Yorkshire Area..79
73 47 Thurcroft Colliery, South Yorkshire Area..80
74 48 Treeton Colliery, South Yorkshire Area...81
75 49 Wath/Manvers Collieries, South Yorkshire Area..82
76 Map 5: Barnsley and Doncaster Areas.........83
77 Grimethorpe Colliery, Barnsley Area, 1992..83
78 50 Barrow Colliery, Barnsley Area...............84
79 51 Bullcliffe Wood Colliery, Barnsley Area..85
80 52 Caphouse Colliery, Barnsley Area...........86
81 53 Denby Grange Colliery, Barnsley Area...87
82 54 Emley Moor Colliery, Barnsley Area.......88
83 55 Grimethorpe Colliery, Barnsley Area......89
84 56 Houghton Main Colliery, Barnsley Area...90
85 57 Redbrook/North Gawber Collieries, Barnsley Area... 91
86 58.1 Royston Drift Mine, Barnsley Area......92
87 58.2 Royston Drift Mine, Barnsley Area......93
88 59 Wooley Colliery, Barnsley Area...............94
89 60 Askern Colliery, Doncaster Area.............95
90 61 Bentley Colliery, Doncaster Area............96
91 62 Brodsworth Colliery, Doncaster Area......97
92 63 Frickley/South Elmshall Colliery, Doncaster Area..98
93 64 Goldthorpe Colliery, Doncaster Area......99
94 65 Hatfield Colliery, Doncaster Area.........100
95 66 Hickleton Colliery, Doncaster Area.......101
96 67 Markham Main Colliery, Doncaster Area..102
97 68 Rossington Colliery, Doncaster Area.....103
98 Map 6: Midlands Coalfields Areas............104
99 Hucknall Colliery, South Nottinghamshire Area, 1955 (English Heritage)...................104
100 69 Arkwright Colliery, North Derbyshire Area...105
101 70 High Moor Colliery, North Derbyshire Area...106
102 71 Markham Colliery, North Derbyshire Area...107
103 72 Pleasley Colliery, North Derbyshire Area...108
104 73 Shirebrook Colliery, North Derbyshire Area...109
105 74 Warsop Main Colliery, North Derbyshire Area...110
106 75 Whitwell Colliery, North Derbyshire Area...111
107 76 Bevercotes Colliery, North Nottinghamshire Area............................112
108 77 Bilsthorpe Colliery, North Nottinghamshire Area............................113
109 78 Clipstone Colliery, North Nottinghamshire Area............................114
110 79 Creswell Colliery, North Nottinghamshire Area...115
111 80 Harworth Colliery, North Nottinghamshire Area............................ 116
112 81 Sherwood/Mansfield Collieries, North Nottinghamshire Area............................117
113 82 Silverhill Colliery, North Nottinghamshire Area...118
114 83 Thoresby Colliery, North Nottinghamshire Area............................119
115 84 Welbeck Colliery, North Nottinghamshire Area...120
116 85 Bentinck Colliery, South Nottinghamshire Area...121
117 86.1 Calverton Colliery, South Nottinghamshire Area............................122
118 86.2 Calverton Colliery, South Nottinghamshire Area............................ 123
119 87.1 Cotgrave Colliery, South Nottinghamshire Area............................124
120 87.2 Cotgrave Colliery, South Nottinghamshire Area............................125
121 88 Gedline Colliery, South Nottinghamshire Area...126
122 89 Hucknall Colliery, South Nottinghamshire Area............................ 127
123 90.1 Newstead and Annesley Collieries, South Nottinghamshire Area................... 128
124 90.2 Newstead and Annesley Collieries, South Nottinghamshire Area................... 129
125 91 Bagworth Colliery, South Midlands Area...130
126 92.1 Daw Mill Colliery, South Midlands Area...131
127 92.2 Daw Mill Colliery, South Midlands Area...132
128 93 Keresley Colliery, South Midlands Area...133
129 Allerton Bywater, North Yorkshire Area...134

Preface

Of the raw materials used by humans coal was the first to be measured in millions of tons. Long before the Industrial Revolution Britain had developed a coal-fuel economy without parallel anywhere. By the end of the seventeenth century annual production stood at about 2.5 million tons, greater than that of her nearest rival in the second decade of the nineteenth century, and by 1800 had reached perhaps ten million tons, equivalent to a ton per head of the population. This astonishing statistic reflects a long period in which numerous industrial processes, ranging from glass-making to non-ferrous metal smelting and brewing, were adapted to take advantage of coal, only partly in response to failing timber supplies. With the growth of iron-smelting based on coked coal instead of charcoal as a fuel from the middle years of the eighteenth century, and the progressive introduction of steam power, itself a significant consumer of coal, Britain was well established as the world's first great coal-fuel economy.

Initially coal was mined where it outcropped or at very shallow depths and was mainly consumed locally, although as early as the sixteenth century London was receiving significant tonnages of 'sea coal' by ship from the Tyne. The ability to raise coal from greater depths became an imperative as demand increased but problems of drainage, ventilation, and, where these had been solved, transport of coal underground were insuperable.

It was the application of steam, in the early eighteenth century, at first for drainage and then for winding, that enabled deep mining to take place. By 1800, of the 2,000 or so steam engines then in use in Britain, at least half were employed at collieries. But despite this, technological change in the coal industry has been largely overlooked by historians. We are only now, as the deep pit becomes a rarity, beginning to fully appreciate the extraordinary impact that the coal industry has had on our national life and to gain some understanding of its profound influence on the landscape.

It was the cost and impracticability of moving coal by road waggon or packhorse that led Huntingdon Beaumont to build some form of 'passage....laid with Railes' from Strelley to Wollaton Lane end near Nottingham in the early 1600s, the first known railway in Britain. By the early 1800s there were several hundred miles of plateway linking coal mines to navigable water, to the Tyne and the sea route to London or the new network of inland navigations spreading throughout the Midlands. The Great Northern Coalfield of Northumberland and Durham and the valleys of South Wales became the engines of new technologies in which the getting and transport of coal were the determining factors. The locomotive engine and the birth of the steam railway, the work of Humphrey Davy, Trevithick, and the Stephensons, were largely predicated on the rising demand for coal on which Britain's growth as an industrial nation was based. Coal came to have a wider significance too, in the development of employment and safety legislation, trades unionism, and the labour movement. The potent symbolism of the Durham Miners' Gala went far beyond the coalfields themselves to touch everyone who lived in what Aneurin Bevan was to call the island '...made of coal and surrounded by fish'.

Even today a substantial proportion of the nation's energy requirements are met by steam and coal in the form of turbo-alternators driven by steam from the coal-fired boilers of some of Europe's largest and most efficient power stations. But elsewhere in the economy the demand for coal has all but disappeared leading to the progressive decline of deep-pit mining.

This survey was undertaken as a record

Changing landscapes: Shireoaks Colliery, South Yorkshire: (above) as it was in 1961; (below) the same site lies in the triangular wedge of wasteland at the centre of the photograph. One building survives, the pithead baths surrounded by what is now the colliery's industrial archaeology. It is possible to make out the two small squares of the capped-off shafts, the contours of the former colliery railway, and the partially water-filled inlet of a wharf on the Chesterfield canal which forms the right hand boundary of the site

of an industrial landscape that was vanishing as the work was in progress. A decade ago there were in the region of 130 deep collieries at work. At the time of writing there were not more than one tenth that number. In response to the need for some form of national survey English Heritage commissioned this aerial study of the deep-mined coal industry from the Royal Commission on the Historical Monuments of England and appropriate text to accompany the photographs from acknowledged experts in the field. The survey is intended to provide a work of reference for all those involved in the management of the industrial heritage. But perhaps it can be viewed too as something of a tribute to the communities who lived and worked in the coalfields and upon whom, for nearly three hundred years, the nation's demand for energy depended.

Neil Cossons

February 1995

Introduction

Background

This aerial survey presents a strategic view of the coal industry. It documents the technological and structural developments of the deep-mined coal industry, the visible surface features which have characterised the landscapes of the coalfields of England and which are at risk of disappearance almost overnight. This is not a study of the social, economic, environmental, political, or the many other issues which are raised by the changes of the last decade. English Heritage is concerned with the recognition and protection of the historic environment, an environment in which the coal industry has been a significant component. Coal has been worked in Britain at least since the Roman period, although the scale of working during early times is thought to have been very limited. Recent discoveries in Leicestershire, however, have shown the existence of complex, underground coal working, dated by dendrochronology to the mid fifteenth century. Since the seventeenth century the industry has been of national importance, though output has declined considerably from its peak in 1913. Thus the recording of the twentieth-century coal industry covers only the most recent period of a long history of coal extraction and industrial history.

At a time of considerable change and challenge for the coal industry, English Heritage has focused on documenting this national industry, once a mainstay of the British economy. The exploitation of coal resources was fundamental to the emergence of Britain as the first modern industrial power but very little recording has been done for the remains of the deep-mined coal industry of the later nineteenth and twentieth century. Indeed the detailed history of mining technology for that period has still to be written.

The poverty of our records for the more modern mining sites, now being swept away through the recent changes in the structure of British Coal, was highlighted at a conference held in Preston in June 1992, organised jointly by the Council for British Archaeology and the National Association of Mining History Organisations. It was clear that even the most basic information about the layout of sites was generally unavailable, and no detailed surveys of buildings, mining structures, and plant had been undertaken for the post-war industry. The speed with which restructuring had overtaken the nationalised industry had not been anticipated and the situation required a swift reaction if essential information was to be collected. English Heritage was able to respond immediately by commissioning the Royal Commission on the Historical Monuments of England (RCHME) to initiate an emergency flying programme aimed at securing basic air photographic cover of the English coalfields, over 150 British Coal-managed sites.

Aerial survey is one of the most effective methods of rapid-response recording for extensive industrial sites and their associated landscapes. The Air Photography Unit of RCHME has considerable experience in this field from its own programmes of archaeological reconnaissance and from the extensive aerial surveys it has carried out, for example, in textile mill and hospital recording projects. The photography commissioned by RCHME for the aerial survey of the coal industry was carried out during September and October 1992. Each colliery was recorded by a series of oblique air photographs taken, wherever possible, from around 1500ft (457m) above ground level. General panchromatic and colour views were specified, from each 'corner' of the site, as well as details of areas of special interest, such as pit head buildings. Over 2000 black and white and colour photographs were taken in the course of this survey and are held in the National Library of Air Photographs in the National Monuments Record Centre, Swindon.

Parkside Colliery, Newton-le-Willows, with its distinctive tall, reinforced-concrete winding towers, was developed from 1957 to 1964 and is shown here in a photograph taken shortly after production started

This book presents the result of that survey, capturing a record of the country's deep-mined coal industry at the moment of its radical restructuring and, on many sites, dismantling. In some instances changes were already well-advanced, as a glance through the photographs in the following pages shows, and only the work of the landscape engineers was documented by the camera. In the short time available for the organisation of the recording the principal objective was to record as fully as possible all working pits and those which had been in operation up until 1986. A number of collieries which had already been closed before that date were included where the flight plan allowed. Similarly, collieries which had already been the subject of clearance were also included where possible. However, no flying was undertaken over the outlying coalfield of Kent, although its last working pit, Betteshanger, closed in 1989, as the collieries of that area were already substantially cleared by the time of the aerial survey.

Inevitably, because of the speed at which the survey target information was assembled, and difficult weather conditions encountered during some of the flying, the photographic coverage is variable. The complexity of mining sites, and the changing, multiple uses of buildings can make the interpretation of air photographs difficult. However, the aim of capturing a comprehensive picture from the air of the morphology of a working industry, its landscape, infrastructure, layout, buildings, and equipment, was fully realised. The survey also documents all the stages of industrial decline and reconstruction, colliery redundancy, demolition, abandonment, clearance, reclamation, and new development. It is a picture of the industrial archaeology of the late twentieth century taking its future shape.

In addition to this aerial record of the English coalfields, the RCHME has undertaken its own ground-based surveys. Over the last five years, more than thirty individual collieries have been photographically recorded by the Emergency Recording staff, selected to represent a cross-section through the development and history of coal mining across the country. In addition, a survey during 1992-3 of over 250 sites was carried out, together with photography of a small number of collieries and their associated communities, designed to capture a 'snap-shot' record of them during a crucial time of change. This record has been published recently as *Images of Industry: Coal* (1994).

The Monuments Protection Programme

Our industrial heritage is under particular pressure. This comes in part from accelerating technological and economic changes which have resulted in the redundancy of many production centres; but also from reclamation schemes, which so often follow hard on the heels of industrial closure, to reinstate land designated as contaminated or 'derelict'. Mine surface structures, processing areas and waste tips have characterised the landscapes of the coalfields of Britain for centuries. Like so much of our past, the presence of these remains begins to have a new significance as they cease to be a familiar sight. As part of its Monuments Protection Programme (MPP), English Heritage is gathering and analysing industrial archaeology information on a systematic basis for the whole country. This is not just for the coal industry but for all classes of extraction and manufacturing industry. By this means, well-informed and clearly documented decisions can be made about the selection of industrial sites for preservation and the best means of protecting them. In accordance with the current MPP strategy for industrial archaeology, the coal industry has been the subject of this structured approach which has as its first stage a synthesis of the development of the industry outlining its surviving features and their characteristics. The brief historical survey on p12 forms a part of this preliminary study.

Industrial monuments present difficulties in terms of the conventional forms of legislative protection, which were not designed for this particular category of conservation need. Extensive areas of industrial 'hinterland' around the production centre can be involved, for which alternative agricultural or recreational uses are not appropriate. Buildings may be variously reused or derelict, suitable, or totally unsuitable for conversion. Partial salvage or reclamation may have continued on an ad hoc basis, and access may be difficult or dangerous. Ultimately, only a few, the best preserved, the most vulnerable, the rare, and the representative can justify preservation, and that can only be judged in the context of the whole picture for the industry. This book presents part of that picture, a view of the deep-mined coal industry of England, as it still existed in the late summer of 1992.

Editorial notes

All the air photographs in this publication are reproduced here with the permission of the Royal Commission on the Historical Monuments of England and are Crown Copyright. These photographs, and over two thousand others taken for the aerial survey, are held in the RCHME's National Library of Air Photographs in the National Monuments Record Centre (Kemble Drive, Swindon SN2 2GZ) where they are available for reference. Copies of photographs may be purchased from the Library and should be ordered quoting the reference number given in the gazetteer index on pp18-20, or in brackets at the end of the caption. The archive photographs illustrating the Introduction and Glossary are part of English Heritage's records for the coal industry and copies are also held in the RCHME's National Monuments Record.

Acknowledgements

The air photography was carried out by Chris Cox of Air Photo Services, Cambridge and the Royal Commission on the Historical Monuments of England. Ian Ayris researched and wrote the descriptive captions for the photographs. The outline of mining history on pp12-15 and the glossary on p134 are taken, in an abbreviated form, from the Monument Protection Programme Step 1 Report for the coal industry commissioned by English Heritage from the Cranstone Consultancy in 1992. The book was coordinated and edited for publication by Gill Chitty.

The coal industry: a brief historical survey of mining technology

by Shane Gould

While there are many modern studies of the coal industry, most of them concentrate on the economic and social aspects of its development. Until recently there has been little detailed coverage of the evolution of mining technology and virtually no study of the surviving evidence in the field. The development of coal mining was not a simple technological progression through time, as new methods and machines were adopted slowly, at different rates in different areas, especially before nationalisation, and it was common for the new to exist alongside the old. There are also considerable difficulties in assigning dates to the earliest, primitive mining sites whose adits, outcroppings, and shaft-mounds cannot be differentiated easily from sites where such simple techniques were maintained until quite recently.

Little is known about the early history of the industry but it is probable that coal was first used in England during the Roman period. There is archaeological evidence for coal-burning from military sites such as Housesteads, and from Corbridge, Chester, and Manchester. Extraction is generally assumed to have been by outcropping, working directly into the surface outcrop of the seam, the fuel then being used for smithing or for domestic purposes. Contemporary documents of this period are few and rarely refer to coal itself. Following the end of Roman administration in Britain in the fifth century AD, the use of coal appears to have ceased altogether for several centuries. There is no mention of it in the Domesday survey of 1086 and it is not until the twelfth century that historical documents referring to coal occur again. Output at this time must have been very small and it is unlikely that there was any regular usage until about 1200.

By the thirteenth century coal was being used in small quantities in most of the areas where it outcropped naturally, especially in Northumberland, Durham, and around the Firth of Forth. Its principal uses were in smithing, for lime-burning, and in the manufacture of salt. Without proper chimneys, it was an unsuitable fuel for the domestic hearth and there was a degree of reluctance in using coal for domestic burning in the towns. Towards the end of the thirteenth century, however, the important trade in 'sea-coals' began to develop between Newcastle and London.

Estimates for total output remain very low until the sixteenth century and it has been assumed, therefore, that the mines and their associated technology were rudimentary. Small quarries, ditches, shallow drifts, and 'bell-pit' workings (shafts that simply bell out at the base to exploit a small part of the seam), for example, have been considered to be typical. More organised, shaft mining was thought to have developed much later. These assumptions must be considerably revised following recent work in Leicestershire (R York and S Warburton, 1991, Digging deep in mining history, *Bulletin of the Association of Industrial Archaeology,* **18**, no 4, pp 1-2), where quite complex underground workings, exposed and destroyed during opencast works near Coleorton, have been dated by dendrochronology to 1450 and 1463.

These workings were of pillar and stall type, in which areas of coal (the stalls) were extracted, leaving intervening pillars of intact coal to support the roof, and were reached by well-timbered vertical shafts. These remains are evidence of far more sophisticated techniques than the 'bell-pit' workings which would have been expected at such an early date, and the evidence points to a strict division of labour including faceworkers, haulagemen, banksmen, and shaftsmen. The Leicestershire discoveries show the existence of deep, organised mines in a period and area where they were not previously known or anticipated. Clearly, preconceived ideas about the rudimentary character of early mining techniques and their surface features need to be reconsidered.

From the second half of the sixteenth century, it is thought that coal output increased considerably because of the growing national shortage of timber stocks. Coal was used more widely for domestic fuel, as chimneys became more common, and also began to be used instead of wood or charcoal for an increasing range of industrial purposes. The transition from a wood- to a coal-based economy took some three hundred years.

Important milestones were the development of coal-fuelled glassmaking in the early seventeenth century, of reverberatory smelting of non-ferrous metals in the late seventeenth century, of coke-fuelled iron smelting in the first half of the eighteenth century, and of the reverberatory puddling furnace for refining pig iron to wrought iron in the late eighteenth century. These developments created an accelerating cycle of consumption. The growing industrial demand for iron placed considerable pressures on coal output and a need to tap larger and, more importantly, deeper coal seams. The solution came in the form of the steam engine which not merely provided the means for the development of the modern mine, but in turn created yet more demand for fuel. The engine was subsequently adopted as a source of motive power to numerous growing industries.

Hem Heath Colliery, Stoke-on-Trent, c 1951: (top) the winding engine house for the shaft dominates the scene. The same building seen below (below,lower centre) is dwarfed by the colliery development of the later 1950s and 1960s

The demand stimulated a corresponding expansion in the scale of colliery workings, both in their numbers and their size. The physical development of sixteenth- to eighteenth-century collieries has yet to be independently studied from field evidence, and the important surviving datable surface and underground features are rare. Simple, hand windlasses for raising the coal may have been replaced by cog and rung or horse gins. Drainage difficulties were overcome by driving adits or by physically drawing the water up the shaft by the simple bucket lift, or rag and chain pump. These were supplemented by waterwheel-powered pumping systems from the sixteenth century onwards, especially in areas where working was being undertaken below adit level.

Headings were driven into the coal from the foot of the shaft and worked on the pillar and stall method, as described above. Longwall working (the extraction of an entire face of coal) was introduced, initially in Shropshire, in the seventeenth century. It gradually replaced pillar and stall working in most areas, though it was never entirely accepted in the north east.

At the beginning of the eighteenth century surface arrangements continued to be relatively simple, consisting of horse gins, open pit tops, storage areas, and perhaps a smithy and one or two small huts. The development of the steam engine changed this picture and, from its introduction in the early eighteenth century, mine surface buildings became more permanent and were dominated by the massive stone-built engine house. The atmospheric (Newcomen-type) engine was developed in the 1710s, gradually becoming more common throughout the century and used chiefly for pumping. The use of steam for winding spread more rapidly after the introduction of the rotary Watt engine in the 1770s.

The scale of many colliery operations was limited by the chronic lack of any transport system, particularly the inland coalfields which were inaccessible to coastal traffic. Considerable difficulties were faced in transporting coals even over the shortest distances and these problems became insurmountable in the winter months. In the north east a system of waggonways, initiated in the 1620s, allowed the expansion of the east coast collieries at increasing distances from the Tyne, exploiting the easily-drained coal seams and dependent on the investment in waggonways for the transport of their output. As the shallow reserves were worked out, in the later eighteenth century production moved to deeper, long-lived collieries close to the lower Tyne, the investment shifting to the deep shafts and steam engines needed to work the heavily-watered lower seams.

Long-distance transport from the inland coalfields only became possible during the canal era of 1760-1830, the mines being connected to canal-side wharves by waggonways. At first, the position of the north east remained virtually unchallenged but growing competition from inland coalfields led to the installation of primitive screening arrangements at the collieries around the Tyne and Wear from about 1770.

The application of steam winding and steam pumping during the eighteenth and nineteenth centuries gave access to greater quantities of coal at increased depths. This in turn stimulated the introduction of improved techniques of shaft winding, coal screening and pit top organisation. Gunpowder was in use from the mid eighteenth century, producing quantities of small coal and slack which, with the coal from poorer seams, required screening before resale. The waste was then tipped onto ever growing spoil heaps that were to become characteristic features of the late nineteenth-century coalfield landscape.

New techniques for underground haulage slowly overcame the problem of the growing distance between pit bottom and the working coal face. In the past wicker 'corves' had simply been dragged or pushed by boys and raised to the surface on freely swinging hemp ropes. In the mid eighteenth century horses were introduced on main roadways, and these were augmented from the 1820s by various forms of steam-driven haulage, which together with fixed guides and cages in the shafts provided an immediate improvement.

Engine houses for twin cylinder, horizontal winding engines, Cornish pumping engine houses, boiler houses, ventilation chimneys, offices, workshops, coking ovens and heapsteads dominated the typically 'modern' mid nineteenth-century colliery. The majority of mines eventually developed integrated rail links which in association with all these features came to dominate colliery design up to and indeed after nationalisation in 1947.

Naturally these surface arrangements never remained entirely static and continued to be modified and improved. Above and below ground electrification gradually replaced redundant steam engines, especially after nationalisation, and pithead baths and canteens became a further addition from 1926. Air furnaces for underground ventilation were supplanted by steam-driven exhausting fans during the 1860s. These were again gradually replaced by electrically-driven centrifugal fans through the twentieth century. Primitive screening equipment was continually updated and is now integrated within complex washing, crushing, and froth flotation units.

In contrast the application of sub-surface mechanisation was remarkably slow, mainly because of a bountiful supply of low-paid labour. As the mines developed, human haulage gave way to animal traction and then steam haulage. However, the physical skills of the traditional miner were maintained at the coal face, where the development of mechanical cutting and loading came surprisingly late. Attempts to introduce the use of compressed air in the late nineteenth-century were largely unsuccessful. Electrically-powered cutters were developed from the 1890s onwards, but mechanisation was slow (only 8% of coal cut in 1913, rising to 61% in 1939). The National Coal Board finally succeeded in the mechanisation of the coal face in the 1950s and automated shears, loaders, collapsible powered supports, and face and main road conveyors signified the arrival of techniques of flow production in this historic industry.

Bibliography

A number of general publications on industrial archaeology contain useful chapters and bibliographies on the coal industry which are good introductions to the subject. Neil Cossons' The *BP Book of Industrial Archaeology* (1987) describes the history of mining technology and has a chapter on the importance and evolution of the steam winding and pumping engine. *Industrial archaeology in Britain* by R A Buchanan (1980) and *Post-medieval archaeology* (1990) by D Crossley also contain useful sections on the industry.

The basic work of reference for the economic development of the industry is the five volume *History of the British Coal Industry* (1984-93), commissioned by the National Coal Board (now British Coal) as an independent and comprehensive history. It comprises *Volume 1: Before 1700* by John Hatcher (1993), *Volume 2: 1700-1830: The Industrial Revolution* by Michael W Flinn, *Volume 3: 1830-1913: Victorian pre-eminence* by Roy Church, *Volume 4: 1913-1946: The political economy of decline* by Barry Supple, and *Volume 5: 1946-1982: The nationalised industry* by William Ashworth. J U Nef's *The rise of the British coal industry*, 2 volumes (1932) remains a classic work on the history of the early industry.

The technology of coal mining is discussed in greater detail by various authors in the *Historical review of coal mining* (published by the Mining Association of Great Britain, *c* 1924) and *Coalmining* by A R Griffin (1971) provides an up-dated analysis with an appended list of surviving sites. *Coal mining: a technological chronology 1700-1950* (1991) by A Hill provides a brief overview.

Ireland Colliery, North Derbyshire: a typical colliery clearance. The wedge-shaped area of the colliery buildings, all demolished but some still showing in clear outline, occupies the centre of this view. The site is surrounded on three sides by spoil tips of coal waste and partially-filled lagoons. Reclamation of the colliery waste is underway

A systematic bibliography, the *Bibliography of the British coal industry* by J Benson, R G Neville, and C H Thompson (1981) has also been published as a companion to the five volume economic history of the industry. Contemporary mining journals also provide a wealth of material. *The Colliery Guardian, The Mining Journal,* and *The Colliery Yearbook and Coal Trades Directory* are particularly useful.

Images of industry: coal by Robin Thorne (1994), and another study of the industry from the Royal Commission on Ancient and Historical Monuments in Wales, *Collieries of Wales: engineering and architecture* (forthcoming), illustrate the recent investigation and recording of collieries in England and Wales by the two Royal Commissions.

Frickley/South Elmsall combined mine, Doncaster Area: 1960s coal preparation plant and skip winding facilities at pithead (14248/15)

Gazetteer

by Ian Ayris

Using the gazetteer

The gazetteer is arranged alphabetically within the ten old National Coal Board administrative areas (p21), and divided into four regional sections for the Western Area, North East Area, Yorkshire, and Midlands coalfields. Nearly a hundred of the collieries included in the survey are illustrated and many more are referred to in the text. Each entry in the gazetteer carries the name of the colliery shown in the photograph, its old National Coal Board Area, and a six-figure Ordnance Survey grid reference. This information is summarised, with the RCHME National Air Photograph Library reference numbers, in the index on p18. An alphabetical listing of the all sites included in the aerial survey will be found on p137.

The descriptions

In the post-war period, production from the nation's coalfields became centred on large-scale mines, partly as a result of nationalisation in 1947 and of major investment programmes directed at a relatively small number of collieries. Modernisation programmes from the 1950s onwards brought similar designs and technology to surface installations throughout the country, particularly in the area of coal preparation. The same configuration of plant, buildings, and operations at the mine surface will be seen on many of the photographs in this gazetteer. The descriptions are therefore designed to chart the history of the nationalised industry and its earlier history. They include general information about the industry, or, on occasion, focus on a unique feature or particular aspect of colliery development which is well-illustrated by a site. A glossary of mining terminology and technical words used in the gazetteer will be found on p134.

Index to gazetteer

Gazetteer number	Colliery name	NCB area	OS grid referenc	RCHME acc no
1	Agecroft Colliery	West	SD 803 014	14371/8
2.1	Bickershaw Colliery	West	SJ 634 997	14372/4
2.2	Bickershaw Colliery	West		14370/8
3	Bold Colliery	West	SJ 550 930	4367/12
4	Cronton Colliery	West	SJ 475 894	14366/2
5	Florence Colliery	West	SJ 915 420	14377/3
6	Golborne Colliery	West	SJ 603 987	14369/3
7	Haig Colliery	West	NX 967 176	14381/3
8	Hem Heath Colliery	West	SJ 885 415	14376/8
9	Lea Hall Colliery	West	SK 058 168	14378/9
10	Parkside Colliery	West	SJ 600 947	14368/8
11	Parsonage Colliery	West	SD 653 009	14372/11
12.1	Silverdale Colliery	West	SJ 815 467	14374/14
12.2	Silverdale Colliery	West		14374/6
13	West Cannock No 5 Colliery	West	SK 008 141	14378/14
14	Ashington Colliery	NE	NZ 262 880	14066/31
15	Bates Colliery	NE	NZ 303 823	14344/3
16	Brenkley Colliery	NE	NZ 235 741	14066/20
17.1	Dawdon Colliery	NE	NZ 435 480	14351/2
17.2	Dawdon Colliery	- -	14351/3	
18.1	Easington Colliery	NE	NZ 438 441	14352/3
18.2	Easington Colliery	- -		14352/9
19	Elemore Colliery	NE	NZ 355 457	14347/8
20	Ellington Colliery	NE	NZ 284 918	14342/7
21	Eppleton Colliery	NE	NZ 363 483	14347/14
22	Houghton Colliery	NE	NZ 338 502	14346/11
23	Hylton and Wearmouth Collieries	NE	NZ 365 584	14345/13
24	Lynemouth Colliery	NE	NZ 297 904	14342/14
25	Murton Colliery	NE	NZ 399 472	14348/10
26.1	Seaham Colliery	NE	NZ 413 496	14349/15
26.2	Seaham Colliery	West		14349/12
27	Shilbottle Colliery	NE	NU 210 080	14066/4
28	South Hetton Colliery/ Hawthorne Combined Mine	NE	NZ 385 453	14348/5
29	Vane Tempest Colliery	NE	NZ 423 503	14350/15
30	Westoe Colliery	NE	NZ 375 670	14344/11

Gazetteer number	Colliery name	NCB area	OS grid ref	RCHME acc no
31	Woodhorn Colliery	NE	NZ 290 885	14343/8
32	Allerton Bywater Colliery	N Yorks	SE 424 277	14356/2
33	Fryston Colliery	N Yorks	SE 458 270	14356/12
34.1	Kellingley Colliery	N Yorks	SE 525 236	14354/11
34.2	Kellingley Colliery	N Yorks	SE 525 236	14354/14
34.3	Kellingley Colliery	N Yorks	SE 525 236	14354/1035 35 35
	Ledston Luck Colliery	N Yorks	SE 431 307	14355/9
36	Prince of Wales Colliery	N Yorks	SE 452 229	14357/12
37	Selby Project/ Gascoigne Wood Drift	N Yorks	SE 523 318	14364/10
38	Selby Project/Shaft Mines	N Yorks -		
38.1	Selby Project/North Selby Mine	N Yorks	SE 640 440	14362/11
38.2	Selby Project/Riccall Mine	N Yorks	SE 640 360	14361/5
38.3	Selby Project/Stillingfleet Mine	N Yorks	SE 600 410	14363/1
38.4	Selby Project/Whitemoor Mine	N Yorks	SE 665 356	14360/1
38.5	Selby Project/Wistow Mine	N Yorks	SE 575 355	14363/11
39	Sharlston Colliery	N Yorks	SE 361 180	14250/9
40	Cortonwood Colliery	S Yorks	SE 408 014	14242/5
41	Dinnington Colliery	S Yorks	SK 518 865	14239/2
42	Kiveton Park Colliery	S Yorks	SK 493 827	14238/6
43	Maltby Colliery	S Yorks	SK 552 925	14241/2
44	Manton Colliery	S Yorks	SK 590 780	14228/5
45	Orgreave Colliery	S Yorks	SK 426 870	14238/9
46	Silverwood Colliery	S Yorks SK	478 941	14241/9
47	Thurcroft Colliery	S Yorks	SK 500 885	14239/10
48	Treeton Colliery	S Yorks	SK 435 878	14238/15
49	Wath/Manvers Collieries	S Yorks	SE 446 010	14240/13
50	Barrow (Barnsley Main) Colliery	Barnsley	SE 367 063	14247/6
51	Bullcliffe Wood Colliery	Barnsley	SE 285 145	14251/7
52	Caphouse Colliery	Barnsley	SE 253 165	14250/13
53	Denby Grange Colliery	Barnsley	SE 268 153	14251/1
54	Emley Moor Colliery	Barnsley	SE 234 131	14251/8
55	Grimethorpe Colliery	Barnsley	SE 405 085	14248/8
56	Houghton Main Colliery	Barnsley	SE 420 062	14247/1
57	Redbrook/North Gawber Colliery	Barnsley	SE 330 008	14247/14
58.1	Royston Drift Mine	Barnsley	SE 375 122	14358/13
58.2	Royston Drift Mine	Barnsley	SE 375 122	14358/12
59	Woolley Colliery	Barnsley	SE 310 110	14252/2

Gazetteer number	Colliery name	NCB area	OS grid ref	RCHME acc no
60	Askern Colliery	Doncaster	SE 561 135	14249/9
61	Bentley Colliery	Doncaster	SE 570 074	14244/4
62	Brodsworth Colliery	Doncaster	SE 525 068	14243/8
63	Frickley/South Elmsall Colliery	Doncaster	SE 445 117	14248/13
64	Goldthorpe Colliery	Doncaster	SE 469 042	14243/2
65	Hatfield Colliery	Doncaster	SE 653 112	14249/2
66	Hickleton Colliery	Doncaster	SE 465 053	14242/13
67	Markham Main Colliery	Doncaster	SE 617 046	14244/8
68	Rossington Colliery	Doncaster	SK 603 984	14244/15
69	Arkwright Colliery	N Derbs	SK 428 702	14230/6
70	High Moor Colliery	N Derbs	SK 468 800	14228/14
71	Markham Colliery	N Derbs	SK 455 718	14230/5
72	Pleasley Colliery	N Derbs	SK 496 645	14232/14
73	Shirebrook Colliery	N Derbs	SK 530 670	14232/8
74	Warsop Main Colliery	N Derbs	SK 546 680	14232/2
75	Whitwell Colliery	N Derbs	SK 535 758	14229/3
76	Bevercotes Colliery	N Notts	SK 695 739	14229/9
77	Bilsthorpe Colliery	N Notts	SK 653 615	14234/1
78	Clipstone Colliery	N Notts	SK 596 632	14233/12
79	Creswell Colliery	N Notts	SK 524 736	14229/14
80	Harworth Colliery	N Notts	SK 625 903	14245/11
81	Sherwood/Mansfield Collieries	N Notts	SK 537 624	14233/3
82	Silverhill Colliery	N Notts	SK 475 620	14234/7
83	Thoresby Colliery	N Notts	SK 635 675	14231/9
84	Welbeck Colliery	N Notts	SK 582 700	14231/4
85	Bentinck Colliery	S Notts	SK 488 550	14235/8
86.1	Calverton Colliery	S Notts	SK 603 500	14236/6
86.2	Calverton Colliery	S Notts	SK 603 500	14236/7
87.1	Cotgrave Colliery	S Notts	SK 651 364	14237/8
87.2	Cotgrave Colliery	S Notts	SK 651 364	14237/12
88	Gedling Colliery	S Notts	SK 612 438	14237/4
89	Hucknall Colliery	S Notts	SK 540 490	14246/1
90.1	Newstead and Annesley Collieries	S Notts	SK 520 532	14235/10
90.2	Newstead Colliery	S Notts	SK 522 533	14235/12
91	Bagworth Colliery	S Midlands	SK 458 068	4746/71
92.1	Daw Mill Colliery	S Midlands	SP 260 900	14379/10
92.2	Daw Mill Colliery	S Midlands	SP 260 900	14379/7
93	Keresley Colliery	S Midlands	SP 320 844	14380/1

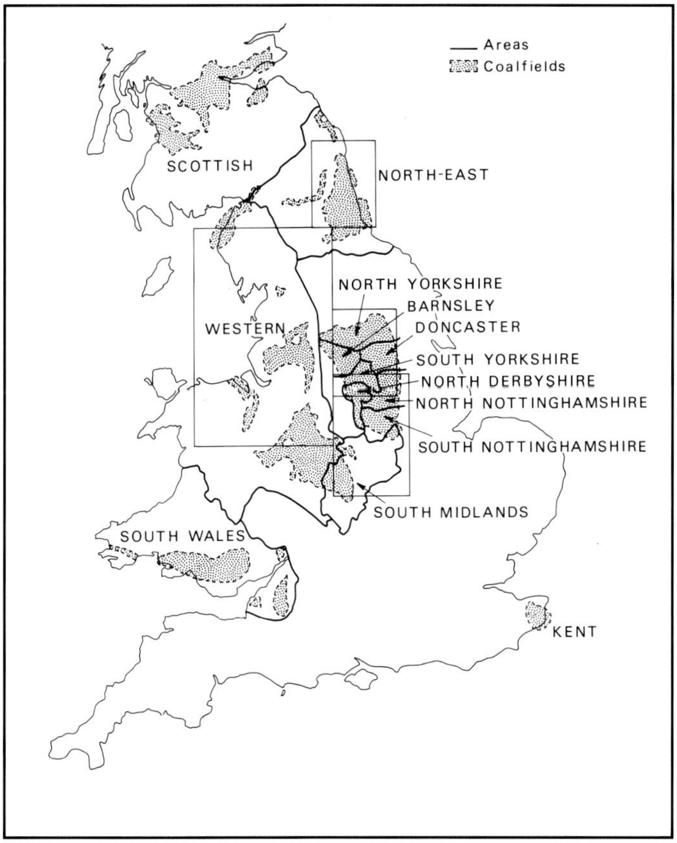

Map 1: NCB areas

Bolsover colliery, North Derbyshire Area, 1 January 1947, the nationalisation of the industry (English Heritage)

Map 2: Western Area

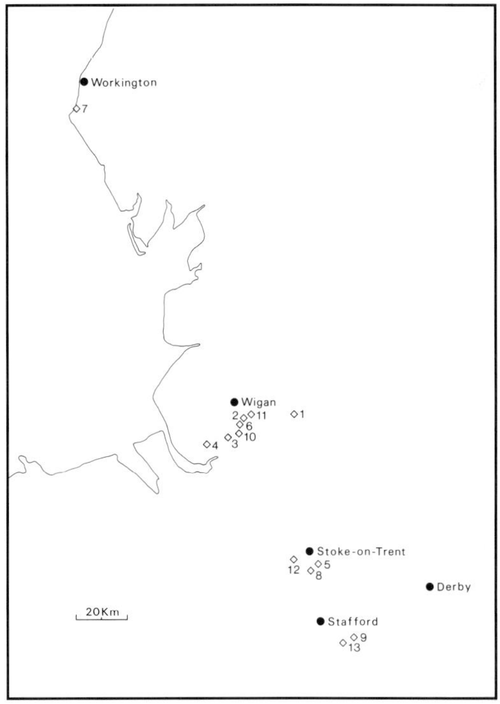

Bickershaw colliery, Western Area, in the 1970s (English Heritage)

1 Agecroft Colliery, Western Area

SD 803 014

Agecroft Colliery (upper centre and right), in Greater Manchester, has now gone but the power station which it provided with coal works on. The 100 acre site (40.5 ha) of the colliery was redeveloped by the National Coal Board from 1953 and was opened in 1960. It was based on the site of an older colliery which had ceased production in 1932. The shafts of the old mine were retained, relined, and deepened. One shaft was provided with skip-winding facilities, one new shaft was sunk, and two concrete winding towers, one 174 feet (53m) and the other 210 feet (64m) high, were built. All the surface buildings were contemporary with the re-opening, the structures from the original mine having been swept away in the 1930s. Half of the output of the mine went directly to the adjacent power station.

2 Bickershaw Colliery, Western Area

SJ 634 997

2.1 Once at the centre of a modern, three-way mining complex with the nearby Golborne (6) and Parsonage (11) Collieries, between Wigan and Warrington, Bickershaw nevertheless retains a large amount of older plant and buildings. Coal from all these mines was raised and treated at Bickershaw. The extensive coal preparation plant, built in 1953 but modified in the mid 1970s when the combine was brought into operation, is in the process of demolition in this picture (centre). The mine itself was started in 1877-81 and worked from four shafts. A modernisation scheme was undertaken in the 1950s. Some of the buildings, garages, workshops and the long baths building (left), date from this time. The waterway which forms part of the perimeter of the mine site is the Leeds-Liverpool Canal (upper centre and right).

2.2 Bickershaw once boasted four headstocks. The No 4 Shaft equipment (lower centre) dates from 1937-8. The adjacent lattice-work arrangement is from the 1920s. Production was concentrated at these two shafts. The No 1 Shaft (upper right) was used for pumping and ventilation. The headstock dates from c1900 and was brought to this site from Maypole Pit, near Wigan. Steam power was in use at this site for over 100 years. Two twin-cylinder, horizontal winders, a steam-driven Walker's compressor, and a bank of eight Lancashire boilers were still in operation in the late 1970s. The Woods winder at the No1 Shaft was thought to date from the sinking of the mine in 1877. The No 3 Shaft was worked by an engine installed by Walker Brothers of Wigan in 1891.

3 Bold Colliery, Western Area

SJ 550 930

Bold Colliery, near St Helens, was sunk in 1876 and completely restructured from 1948 to 1956. Further improvements to the winding apparatus were undertaken in 1976-7 at what was then one of the largest collieries in the north west of England. The mine is now closed but the surviving buildings give an opportunity to look at the use and layout of a typical group of colliery ancillary buildings. The buildings date from 1948-56. Nearest the camera (lower left) are the colliery offices and the lamproom. A covered corridor links these with the pit head baths building. Within the same complex were the first-aid room, the bus waiting room, pay hall, canteen, and kitchen. The building also had a covered bicycle park. The three other buildings on the site are the general manager's office (centre), the chemical laboratory (right), and the sampling laboratory (right centre).

4 Cronton Colliery, Western Area

SJ 475 894

The development of Cronton Colliery, eight miles east of Liverpool, was interrupted by the First World War. The mine was started in 1914 but was not fully open until 1922. The mine is now closed and the buildings demolished with the exception of the former garage for the landsale lorries (lower right centre). The buildings of the colliery surface were situated north of (above) the lagoon at the centre of the photograph. Until its closure the colliery had changed little from its original appearance. The winding engine houses, workshops, offices, loco shed, lamproom, screens, and a small pump house were all contemporary with the sinking, as were the headstocks, the brick air-lock at the upcast shaft, and the Walker's Indestructable fan and fan house. In 1934 pit head baths were added to the site. The winders were both electric, one of which pre-dated the mine, being built by Siemens Brothers in 1903.

5 Florence Colliery, Western Area

SJ 915 420

It was not uncommon in the nineteenth century for colliery owners to name individual pits and occasionally whole collieries after their wives or children. Florence Colliery is one of the few remaining examples of this tradition. The mine, at Longton, Stoke on Trent, was started in 1874 but a £7 million reorganisation scheme in the years from 1950 to 1964 saw the complete renewal of the surface structures. Ten years later it was merged with Hem Heath Colliery (8) to form a giant combined mine. The No 1 Shaft (extreme right) is topped by a concrete tower, whereas the No 2 pit (right centre) and the No 3 pit (lower centre) have steel headstocks. The heapstead of the No 3 pit is enclosed with the flat-roofed, square winder house at its south-western end (lower left). Next to the No 3 winder stands a small boiler house (left) and beyond that the colliery methane plant (left).

6 Golborne Colliery, Western Area

SJ 603 987

Another deserted site in what was once the second most productive coalfield in the country. Golborne Colliery, near Warrington, was sunk in 1882 and eventually became part of the Bickershaw/Parsonage Complex (2). When the combine was formed a number of newly-redundant buildings were removed including the coal preparation plant. At the same time much of the rail network was lifted, the large reservoir filled in, and the boiler house, with its 500 foot (305m) brick chimney demolished. The mine continued to be used for man-riding and ventilation. The shafts were situated near the area where a small treatment plant has been erected to rework the mine waste (left centre).

7 Haig Colliery, Western Area

NX 967 176

An area of considerable industrial archaeological and historical interest, detailed records of mining near this Whitehaven site go back to the beginning of the eighteenth century. The remains of an eighteenth-century engine house stand above the coastal cliffs. The nearby Corkickle Brake Incline, once connected to the Haig Colliery, was one of the last working surface railways, inclined with a standard guage. Deep coal mining in Cumbria, however, came to an end with the closure of Haig Colliery. New units now surround the two winding houses, the compressor house, and the No 5 Shaft heapstead which remain here (centre). The Haig Colliery was started in 1914 and the extant buildings probably date from this time. The winding engines were steam-powered, double cylinder, horizontal engines installed by Bever, Darling & Co of Bradford. The surviving headstock served the men and materials shaft of the mine. The site of the production shaft is marked by the square concrete pad adjacent to the winding house nearest the camera (lower centre).

8 Hem Heath Colliery, Western Area

SJ 885 415

Known as the Big 'A' pit because of its distinctive headgear (centre), Hem Heath Colliery, south west of Stoke-on-Trent, is a modern colliery rebuilt on the site of an earlier pit. Completely modernised in the 1950s and 1960s, Hem Heath became north Staffordshire's biggest producing mine when it was linked underground to Florence Colliery (5) and a new drift was driven in 1977-8. It had begun as an outlying shaft for Stafford Colliery, sunk in 1924-6 but was developed as an independent mine by the National Coal Board from 1950, eventually beginning the winding of coal in 1956. The formation of the combine, known as the Trentham Project, also saw the building of a new coal preparation plant (extreme left) which came into use in 1978. The old preparation plant was retained (upper centre and left centre). A skip winder was installed at the No1 Shaft (lower right). The drift from which most production emanates is on the lower left of the photograph.

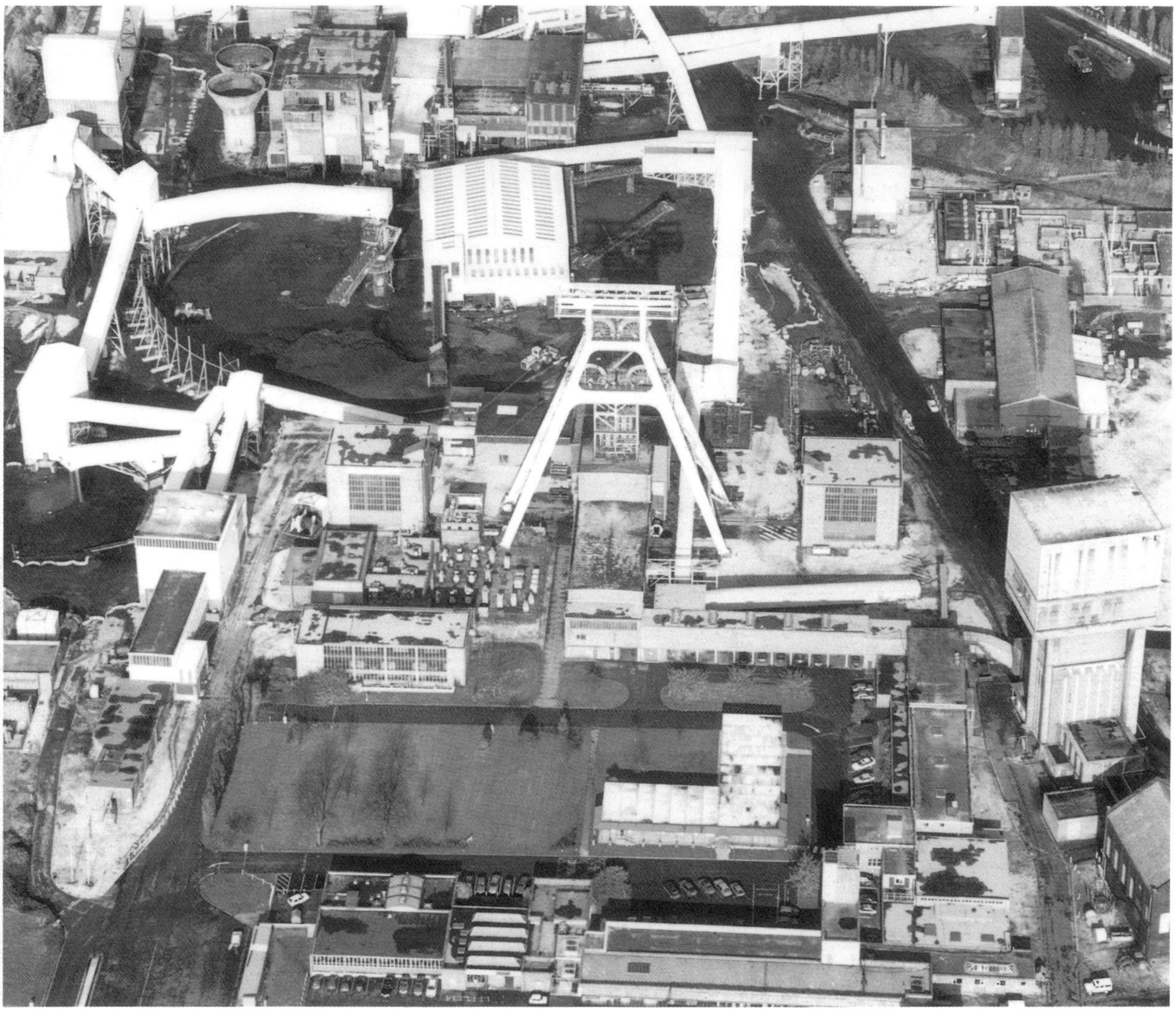

11 Parsonage Colliery, Western Area

SD 653 009

The surface area of Parsonage Colliery, sunk in the period from 1913 to 1920 was substantially reduced when the Bickershaw (2), Golborne (6), and Parsonage Combine was completed in 1976. In 1992 it was reduced even further following the closure of the mine. The camera has caught the demolition in progress. At the centre is the tall brick chimney which once served the boiler house where eleven boilers powered colliery machinery until the 1980s. The No 1 Shaft winder was a double-cylinder horizontal engine installed in 1923. Winding at the No 2 Shaft was electrified in the colliery's later life. Part of the No 2 winder is still visible (lower right). The winding drum of the No 1 engine can also be seen immediately right of the chimney base. Around the site other buildings are in varying stages of demolition, notably the decaying remains of the 1920s power house (lower left centre). Around the perimeter of the site are former offices (upper centre), cycle sheds (upper right centre), the pit head baths (upper right), lamproom (right), and the canteen and group offices (extreme right).

12 Silverdale Colliery, Western Area

SJ 815 467

12.1 Silverdale is an extensive Newcastle-under-Lyme colliery massively modernised in the late 1970s. While the colliery has the usual paraphenalia involved in the preparation and distribution of coal, this view has been chosen to illustrate the lack of structures required at the mouth of the modern surface drift. The No 3 drift portal and fan house are in the foreground (lower centre and lower left) while the No 1 drift emerges at the building opposite (centre). Coal has been mined in and around Silverdale since at least the mid eighteenth century. The present colliery was based on four shafts, the first of which was sunk in 1870.

12.2 Silverdale led the way in the development of water treatment equipment at a coal mining site. The plant (centre) treats mine water to a level where it can be either suitably discharged or reused within the colliery. This is a particular problem in this area where mine water contains high iron content. The treatment plant was developed partly because of the lack of space for the provision of the more commonly found lagoons.

13 West Cannock No 5 Colliery, Western Area

SK 008 141

New units encroach upon the land which formerly held the winding houses and heapstead buildings of West Cannock No 5 Colliery, Staffordshire (upper left and right). The two shafts of the mine had been started in 1914. A number of new surface facilities were added to the site in 1957 including the offices which, like a number of other former mine buildings, are still in use here. The baths, medical centre, time office, and control room complex have been retained (lower left centre), as have the single-storey lamproom, rescue room, and office building (centre), the workshops and stores (lower right), and the colliery garage (lower right).

Map 3: North East Area

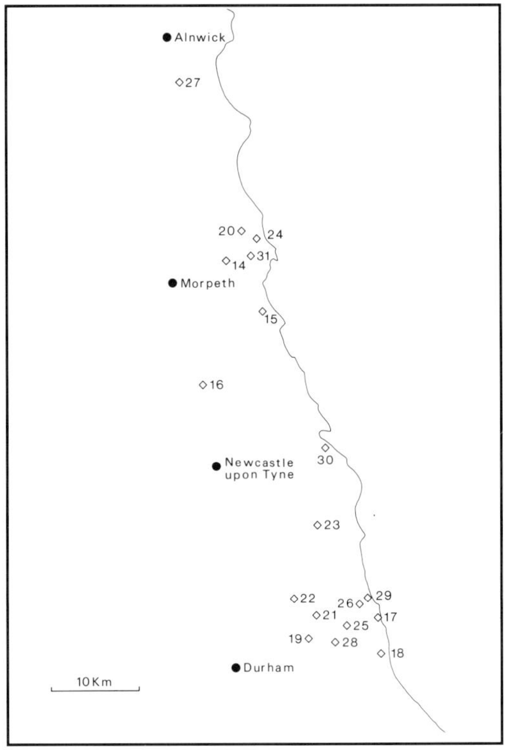

Eppleton colliery, North East Area: Lindsay shaft concrete, air lock tower and winder house, 1970s (English Heritage)

14 Ashington Colliery, North East Area

NZ 262 880

The cleared site of Ashington Colliery does not do justice to what was the most important and influential centre of mining activity north of the River Tyne. Ashington, at its height, was a mighty complex of six shafts, towering chimneys, gasworks, brickworks, a power station, and extensive workshops. The origins of the mine date from 1849 but the oldest shaft on this site dated from 1867. The colliery grew to be one of the largest in the county. Its owners, the Ashington Coal Company, controlled a series of mines and were responsible for the growth of the country's largest mining town where row upon row of miners' houses spread outward from the entrance to the pit yard. The colliery site is now empty but the legacy lives on in the long parallel terraces stretching into the distance. The colliery site itself has now been partly used for a bypass and newly-laid roads prepare the site for future use. In the early 1960s there were no less than 241 separate building uses on the colliery and the nearby area workshops site.

15 Bates Colliery, North East Area

NZ 303 823

The town of Blyth holds a significant position in the development of coal mining in south east Northumberland as the home of some of the earliest deep shaft mines. Bates Colliery, however, dates from 1932, although one shaft on the site was sunk in 1904. The two capped-off shafts of the closed mine can be seen as octagonal features on the right of the picture. The riverside location allowed the coal, mined from beneath the North Sea, to be loaded via staiths into ships for transfer to power stations in the Thames area and in the south of England. The twin shipping hoists are still operative at the riverside but some river facilities formerly used by the colliery are redundant, for example the stone shipping jetty on the left of the photograph.

16 Brenkley Colliery, North East Area

NZ 235 741

Brenkley Drift was the last producing element of a long-worked site. Seaton Burn Colliery was started in 1844 and worked via shafts and later drifts until the driving of Brenkley Drift by the National Coal Board. The drift, which started production in 1955, was isolated from the original colliery site and the two were connected by an inclined railway. The route of the line fell steeply from Seaton Burn, under the main A1 trunk road from Newcastle to Morpeth and down to the drift mouth. The line of the incline from Seaton Burn enters the the picture from the lower centre and travels to the site of the former drift mouth, now represented by a concrete pad, visible to the left of the centre of the photograph. From there a trackway continued down the slope to an upcast shaft, winder house, and fan house situated near the trees at the upper left of the picture.

17 Dawdon Colliery, North East Area

NZ 435 480

17.1 Dawdon Colliery once perched upon the shallow cliffs above the North Sea coast on the outskirts of the port of Seaham Harbour. Sunk from 1900 to 1907, the mine underwent a major transformation in 1962 when a £4 million investment programme was carried out. Formerly one of County Durham's million ton coastal mines working the undersea seams, the site has now been cleared. An octagonal shaft cap can be seen at the centre of the photograph and the route of the colliery railway complex can be seen curving along the coastline in front of Noses Point spoil heap. This last feature was a coastal tipping point where two tippers spilt their load from a height into the sea, helping to create the coal-covered beaches of this stretch of the County Durham coast.

17.2 The northern end of the former Dawdon Colliery site is just visible on the lower right hand edge of the picture, which focuses upon the fascinating port of Seaham Harbour, formerly the home of Seaham (26) and Vane Tempest (29) collieries. The port, as well as containing a number of historic industrial structures, is of particular interest in the development of the Durham Coalfield. It was purpose-built by the hugely influential landowner and coalowner, Lord Londonderry, to handle coals from his mines in opposition to the stranglehold maintained on the early and mid nineteenth-century coal trade by the river authorities on the rivers Tyne and Wear.

18 Easington Colliery, North East Area

NZ 438 441

18.1 One of County Durham's coastal pits, Easington colliery is the focus for the mining town of Easington, one of the largest, colliery-based communities in the former Great Northern Coalfield. The colliery itself was opened in 1910, the shafts having taken ten years to complete. The mine was deepened as part of a major reconstruction programme completed in 1958. The site is a mixture of old and new. The heapstead building (centre) surrounding the north and south shaft mouths dates from earlier in the century and may even be contemporary with the opening of the mine, whereas the tall and narrow rapid loading bunker (upper right) is of modern construction. Many of the workshops are older buildings re-roofed in modern materials, for example the fitting shop adjacent to Station Road (extreme right).

18.2 At the centre stand the elegantly designed pit head baths of 1937. One of the earliest references to the building of pit head baths in the North East Coalfield was in April 1855 when Cramlington Colliery owners were reported to have opened 'warm baths and showers at the colliery for their men'. However, pit head baths are more of a twentieth-century phenomenon, a large number of baths being built following the Mining Industry Act of 1926. Many architecturally interesting buildings were erected by the Miners' Welfare Committee from 1933 onwards in Northumberland and Durham, a number designed by F G Frizzell. Approximately thirty such buildings were erected in Durham in this period, all brick-built and characterised by the dominant display of chimney and water tank. Particularly impressive examples were built at Elemore (19), a surviving listed building, and here at Easington.

25 Murton Colliery, North East Area

NZ 399 472

A casualty of the early 1990s, the buildings of Murton Colliery had already been removed by the time of this photograph, except for the two sub-station buildings (right) and Koepe Winder (centre) which is a listed building. The middle pit heapstead, on the other hand, has been swept away leaving the capped-off shaft visible at lower extreme left. The mine had been one of the most important in County Durham and became the major component of the Hawthorn Combined Mine (28) which brought together the output from both Murton and Eppleton (21) at a central shaft sunk from 1952 to 1958. Elemore Colliery (19) had also been part of this operation. Following this reorganisation stone only was raised at Murton. The Koepe Winder had been installed in 1923 and as such was only the second use of an electrically powered friction winder in the North East.

26 Seaham Colliery, North East Area

NZ 413 496

26.1 Seaham, south of Sunderland, originally one of Lord Londonderry's collieries, started in the 1840s as two mines, Seaham and Seaton, and was at one time one of the deepest mines in the coalfield. Its origins explain the unusual distance between the two shafts. On this picture only the coal handling No 1 shaft and headgear can be seen (left). The operation of the surface installations can be easily followed. The coal is wound up the shaft by the winding engine (upper left), passes to the screens where the coal is filtered (centre), then on to a tall cylindrical coal bunker (right centre), then to the washery with its clarifiers and settling tanks (centre) before being outloaded. Seaham was the last mine in the country to work a standard gauge, rope incline-controlled by a stationary engine in the pit yard. The line of the former incline railway can be seen entering at extreme left and running diagonally across the picture to the lower extreme right. The feature at the lower extreme left which has the appearance of a series of large connected pipes is an underground locomotive training unit.

26.2 The principal point of interest amongst the site structures at Seaham was the tower of No 3 heapstead (centre). Thought to contain part of the old vertical winding engine house, parts of the structure date from the sinking of the mine. Monumental cut red sandstone blocks formed the base of the building, with brickwork above. The structure over the years was much altered and disfigured with patches of infill brickwork and cement rendering. The rarity of this type of winding house made even these limited remains of particular interest.

27 Shilbottle Colliery, North East Area

NU 210 080

'What kind of coal does the devil burn in Hades ?'
'Shilbottle, because it leaves little or no ash for his Satanic Majesty to shift.'

This saying from the 1920s indicates the renowned reputation of Shilbottle coal. The colliery continued working difficult and low seams by hand tools until 1982. On the ending of production, the shaft was kept open as an air shaft and for man-riding for nearby Whittle. The most northerly National Coal Board mine in England, lying a few miles south of Alnwick, the site installations dated mostly from the sinking of a new shaft in 1925, but the history of mining here is much older. Shilbottle is thought to have been the site of a Newcomen steam engine in 1769, to have had a metal railroad in use by 1809, but conversely is known to have still been using wicker corves for carrying coal as late as 1864. In the modern era the coal production of the mine came to the surface at Whittle Drift.

Perhaps because of the relatively isolated and rural setting of the mine no reclamation scheme has swept away the visual evidence. Even after ten years the individual building plots can be distinguished. The substantial house (right) was the colliery manager's house.

28 South Hetton Colliery/Hawthorn Combined Mine, North East Area

NZ 385 453

The Hawthorn mine, County Durham, was the North East Area's most ambitious mining project of the 1950s bringing together the production of a number of neighbouring collieries at one shaft. Situated adjacent to, and dwarfing the early nineteenth-century South Hetton Colliery, the Hawthorn mine site has now been cleared of all but a coal bunker (centre) and an electricity sub-station (upper centre). Part of what was a million gallon-capacity lagoon can also be seen (right). The shaft itself, 1,491 feet (305m) deep and once disgorging a million tons per annum, was situated beyond the bunker in the centre of the photograph. The former South Hetton Colliery, historically notable for the introduction of underground iron tubs which led to the development of shaft cages, lay to the west of this site.

29 Vane Tempest Colliery, North East Area

NZ 423 503

Vane Tempest was the last of the great Durham coastal collieries to be sunk. The name of the mine originated from the family name of the wife of the first Lord Londonderry. The shafts were the deepest sinkings in the coalfield. Both the Vane Shaft (centre) and the Tempest Shaft (upper left centre) were completed in 1928. The buildings date principally from the 1930s, despite the reorganisation and mechanisation scheme of the early 1960s. Both winding engine houses, which can be seen at either end of the long heapstead, were built at this later time (centre and upper left). The colliery retained its original screens or picking belts (lower centre) but benefited from a new twin bunker (lower left). Among the other buildings which survived the reorganisation were the huge compressor house (upper centre) and the loco shed (lower right centre). The architecture of the loco shed combines the usual features of early 1930s colliery buildings with a hint of the railway, a high brick-built single-storey building, arch-headed window and door openings with decorative keystones, high circular windows on the gable ends, and large wooden double doors. At the centre of the photograph, the truncated remains of a chimney serve the boiler house. With the changeover to electric winding the steam requirements of the site were vastly reduced, but two boilers were retained to supply the pit head baths and heating inside some buildings.

30 Westoe Colliery, North East Area

NZ 375 670

The early 1960s enclosed winding tower of the Crown Shaft (right) is a local landmark in the town of South Shields at the mouth of the Tyne. Begun in 1909, as a man-riding shaft for the nearby St Hilda Colliery, Westoe became a producing mine when the old colliery closed in 1940 partly as a result of the loss of markets resulting from the outbreak of the Second World War. A major reconstruction scheme commenced in 1957 to work the undersea reserves. Until c1990 the site was serviced by the Harton Electric Railway which took initially coal but later only stone to staiths on the Tyne. The use of an electrically operated railway in a coalfield was highly unusual. The system's origins lay with the colliery's first owners, the partly German-owned Harton Coal Company, this system being common in the German coalfields.

The site layout is unusual in having the coal preparation plant (left centre) between the two shafts. This view, from the south west, gives a good view of the slurry cones and thickeners which form part of the preparation plant (lower centre). In the foreground (lower left) is the Westoe Shaft with its attached substation, compressor room, electrical stores, and generator room in probably the most ornate building on a colliery site in the north east of England.

31 Woodhorn Colliery, North East Area

NZ 290 885

Now preserved as a museum, Woodhorn Colliery on the eastern outskirts of Ashington was the last colliery in the north east to use steam-powered winders. The museum site retains the principal buildings of the mine. As coal production from the colliery was raised at Ashington (14), Woodhorn had only a small range of structures. The site is focused upon a fine pair of turn-of-the-century winding engine houses (centre) with adjoining heapstead structures, one of which at the time of the photograph is shrouded in sheeted scaffolding. A traditional single-storey range of workshops and stables stands upper left to upper centre. The offices are in the upper extreme left corner of the picture and the fan house can be seen at lower centre. The lack of coal preparation installations, the clearing of some buildings, and the general tidying of the site has created a neat and ordered appearance not always typical of working sites. With the loss of so many collieries in this once mighty, and internationally famous coal-producing area, the preservation of this fine set of mine buildings is very welcome.

34.2 The type of coal preparation equipment introduced at many sites in the early 1960s was provided at Kellingley from its inception. The main washery (left), the easily identifiable circular slurry thickeners (extreme left), the covered conveyors, and the rapid loading bunker (upper left) became common sights at the country's pit heads.

34.3 With the ability to plan from the beginning the baths, canteen, medical facilities, and offices were situated adjacent to, but separate from, the shaft-top area. An overhead walkway gave direct access from these facilities to the shaft-top (upper left).

35 Ledston Luck Colliery, North Yorkshire Area

SE 431 307

Sandwiched between the pit head baths building (left) and modern workshops (upper right) are the unusual buildings of the former Ledston Luck Colliery. Situated at Kippax, near Leeds, the mine was opened in 1911 and sunk to a depth of 213 yards (195m). The three surviving buildings date from the opening of this small colliery which was one of the first mines in the Yorkshire area to install electric winding equipment. While on many other sites equipment was introduced into existing, formerly steam-powered, winding houses, here the engine house was built specifically for electrically powered plant. The engine house design, particularly for a colliery site, is highly ornate.

36 Prince of Wales Colliery, North Yorkshire Area

SE 452 229

The scene of another major investment programme by the National Coal Board in the late 1970s, production at the Prince of Wales Colliery came to concentrate on high seams worked by surface drifts (left). Originally sunk in 1872, the mine was worked by shafts to a depth of 716 yards (654m). The modern redevelopment scheme saw not only the introduction of two drifts but also a new coal preparation plant. The use of steel frames and profiled cladding allowed large scale structures to be built at relatively low cost. The scale of the modern buildings (centre) is in marked contrast to that of the older small workshops near the drift mouth (left).

39 Sharlston Colliery, North Yorkshire Area

SE 361 180

Sharlston Colliery, near Wakefield, was started in 1873 and worked to a depth of 420 yards (384m) to the Kent Thick and Low Barnsley Seams. Production in the late 1970s had reached 780,000 tons per year, the mine employing 1210 men. The nearby railway line (upper left) carried the mine's coal which was loaded by a rapid loading bunker into liner trains. By this method 1000 ton trains were loaded while still on the move. Once again the abandoned site forms a typical landscape feature with its lagoons, sprawling areas of waste, wedge-shaped working area, and the footprints of its buildings still visible.

40 Cortonwood Colliery, South Yorkshire Area

SE 408 014

A long-range view of the former Cortonwood Colliery, between Barnsley and Doncaster, places the mine in its environment. Industry, agriculture, and housing combine to give a picture of the local economy. The spoil and dirt heaps of the former mine are prominent (centre). The colliery buildings stood at the base of the heap; a collection of vehicles are now assembled on the site. The line of the mineral railway cuts across the photograph and running parallel to it is part of the Sheffield and South Yorkshire Navigation Canal. The rectangular sludge beds and the circular filter bed of the local sewage works straddle the area between the canal and Beech House Road which passes Lundhill Farm (extreme right). The colliery was part of the local economy for over a hundred years, having been sunk in 1873.

41 Dinnington Colliery, South Yorkshire Area

SK 518 865

Although mostly dismantled, several elements of the surface arrangement of this colliery, north-west of Worksop, still stand. On the north (lower) side of Outgang Lane, among tall conifers, are the baths and medical centre. At the pit yard entrance, near the road junction at the centre of the photograph, stands an office building and next to it the rectangular shape of the former colliery reservoir. The two winding engine houses for the No 1 Downcast Shaft and the No 2 Upcast Shaft stand parallel to each other in the centre of the former complex (centre). The shafts were situated to the left of the winders. A bunker from the coal preparation plant is at the exact centre of the picture and the rectangular but round-ended shape of two tailings lagoons lie above that. The former South Yorkshire County Council reclaimed two area of tipping while the colliery was still in operation (extreme left, left, and upper left centre).

42 Kiveton Park Colliery, South Yorkshire Area

SK 493 827

Kiveton Park lies between Sheffield and Worksop and is now served by a drift, but its history dates back to the shaft-worked mine started in 1866-7. The surface installations contained a stone-built engine house and a large brick chimney, both now gone, along with the colliery offices which had been converted from a Victorian school. The drift (centre) feeds a large preparation plant (upper centre) at which stands a set of waiting coal trucks.

43 Maltby Colliery, South Yorkshire Area

SK 552 925

A large amount of investment, modernisation, and reorganisation was undertaken at Maltby, east of Rotherham, in the later 1980s. This view from the south west shows the new arrangement of the No 3 Shaft and coal preparation plant on the right hand side of the complex. The older part of the site lies on the left. The No 1 Shaft (centre) is served by a brick winding engine house built in 1911. Attached to the side of the winder house is the power house. A new workshops building (upper centre) lies north of the No 2 Shaft (centre) but the older single-storey blacksmiths' and fitters' shop has been retained nearby (upper centre). This, like the No 1 engine house, dates from the early years of the mine which was sunk in 1907-10. The pithead baths, medical centre, canteen, and administration block are at the centre of the picture, adjacent to the car park. This block also contains an elevated lamproom. Direct access to the No 1 Shaft cages is given via gantries from the lamproom (centre).

44 Manton Colliery , South Yorkshire Area

SK 590 780

This north-facing view of the turn-of-the-century Manton mine allows identification of all the principal buildings on the surface. Nearest the camera (lower extreme right) is the No 4 Upcast Shaft enclosed winding tower and fanhouse with its typical forked chamber arrangement beyond it. Gantries lead off to the coal preparation plant at the top of the complex (right centre). The small No 1 Shaft and winder stand, almost camouflaged, at the base of the plant (right). To the north of the preparation plant lie the baths, the offices and canteen and below them the darker roof of the lamp cabin stands out (centre). To the south of the preparation plant is the headstock of the No 2 Downcast Shaft with its pre-First World War winding engine house (right centre). To the left of this are the fitters' and electricians' shop and next to them a further workshop building (centre), all dating from the same period as the winding house. This view also allows easy identification of some of the less prominent buildings found on colliery sites. The colliery stores stand next to the gantry leading to the dirt bunker (left centre), the garage for the dumpers and heavy plant is in the foreground (lower centre), the sub-station stands fronting on to the access road (centre), and to the right of it is the colliery methane house (right centre).

47 Thurcroft Colliery, South Yorkshire Area

SK 500 885

The camera has found demolition taking place at this small South Yorkshire colliery site. The gantries to the box-like coal preparation structure (centre) are being dismantled. The smaller modern structure parallel to the preparation plant is a Bradford Breaker, to where materials are fed via the covered conveyor from the heapstead of the downcast shaft. The headgear of this shaft can be seen at the centre of the picture. The downcast, shaft winding engine house, adjacent to the shaft, the power house standing at right angles to it, and the typical long row of single storey colliery workshops nearby (centre) probably all date from the sinking of the mine in 1909-12. The pithead baths stand at the south end of the site at the junction of the site roads (right). The electricians' shop is situated behind the baths (right). Perhaps the most prominent feature is the pit reservoir (lower left centre). The use of water at collieries could be very large indeed, particularly when steam power was in use.

48 Treeton Colliery, South Yorkshire Area

SK 435 878

The demolition of the buildings at Treeton, south of Rotherham, has perhaps been carried out by a railway enthusiast, for both the old and the new loco sheds have survived the clearance. The old loco shed (left centre) stands in isolation, the workshops and heapstead buildings which were its companions have been swept away. The modern loco garage with its curved roof at the western end of the site (right centre) is at the head of the drift mouth, the connecting line to which can still be seen (centre). The largest remaining structure is the substantial office block (centre) which stood guarding the entrance to the yard from the road, Pit Lane. The colliery was linked underground to Orgreave mine (45) and drew out coal from both sets of workings. Coal preparation was carried out at Orgreave, allowing the equipment at Treeton to be removed in the early 1980s. The colliery dated from 1877.

49 Wath/Manvers Collieries, South Yorkshire Area

SE 446 010

Working for over a hundred years, Wath Colliery began in 1876. An original winding house survived on site almost until the closure of the mine. Now all the buildings have gone. The site was at the centre of the photograph. The road access was by Moor Road which enters the picture from the extreme right and runs into the centre of the former mine surface. The car park and the footprint of the canteen building are visible at the end of the road. The shafts of the mine lay in the area next to the former railway (centre). Some reworking of the site is taking place, coal treatment equipment stands at the centre of the photograph. On the right and lower right are the parallel marks of the once extensive sidings with former railway buildings beside (right). In the distance, beyond the circular filter beds of the sewage works, is the edge of the deserted Manvers Mine site (upper right). Manvers was the home of the Area and Regional Offices for the National Coal Board's South Yorkshire area and had been the location of a mine since 1875-7.

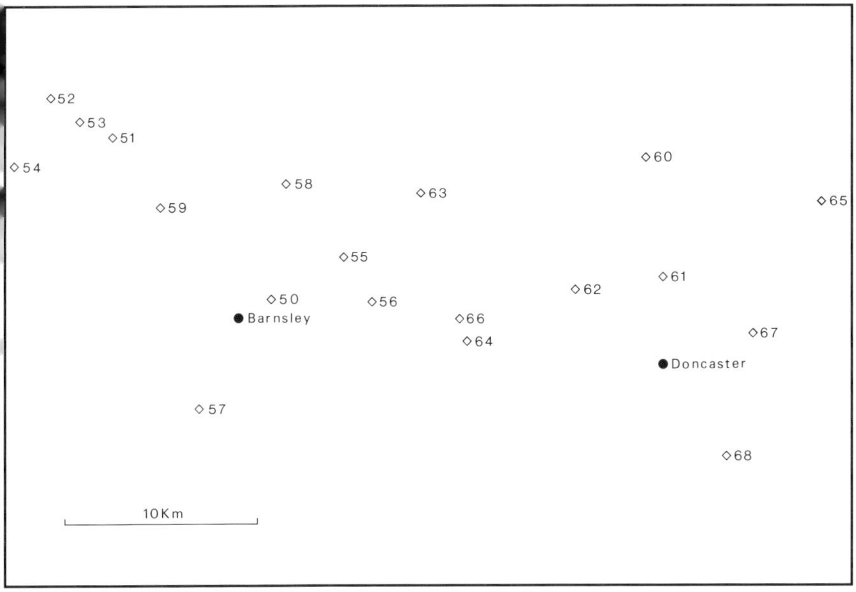

Map 5: Barnsley and Doncaster Areas

Grimethorpe colliery, Barnsley Area, coal preparation area, 1992 (14248/4)

50 Barrow Colliery, Barnsley Area

SE 367 063

The Barrow or Barnsley Main Colliery was based on three shafts sunk in 1875 of which two brought coal from 300-400 yards (274-365m) below ground while the third wound men and materials. The site (centre), near Worsbrough Bridge, south of Barnsley, is now virtually cleared but the town of Barnsley is still inextricably linked with coal mining. From the miner on the civic coat of arms to the nickname of the towns football team, the Colliers, the town's history is steeped in the coal industry. It lay at the centre of the once prosperous and vibrant Barnsley coalfield where most of the mines worked the Barnsley Seam until dwindling reserves led to the search for, and exploitation of, deeper seams.

51 Bullcliffe Wood Colliery, Barnsley Area

SE 285 145

Perhaps only in the North East Area was there a decline in the number of coal mines in the 1980s as steep as that in the Barnsley area. In 1978 twenty mines formed the Barnsley administrative area of the National Coal Board. By 1983 a handful had closed leaving fifteen at work. This site is one of the nine of those fifteen where the colliery had closed and the site had been cleared by the time of this aerial photography survey in 1992. Of the remainder, one is preserved as a museum and further closures have now taken place since these pictures were taken. The site of Bullcliffe Wood Colliery, to the north west of Barnsley, is here deserted but heavy machinery is reforming the waste heap. The mine itself was opened in 1927 and was latterly worked from drift mines established in 1956. Two of the drifts dipped at a gradient of 1:4, one being 405 yards (370m) long and the other 469 yards (429m) long. The former was equipped with a 36 inch (914mm) conveyor which delivered directly into a bunker; the latter was used for ventilation and as access for the workforce.

52 Caphouse Colliery, Barnsley Area

SE 253 165

Appropriately Caphouse Colliery is now preserved as a coal mining museum recording the considerable history of the industry in an area where mining has been of prime importance. Coal extraction at Caphouse has origins in the eighteenth century but the present complex is thought to have grown from early nineteenth-century beginnings. It was a small-scale operation and retains a number of buildings and installations from the late Victorian period, including a wooden headstock and a small stone-built engine house with a steam winding engine and a pair of Lancashire boilers. The mine in its later years had worked via a drift driven in 1974. The site is situated six miles from Wakefield on the A642 road to Huddersfield.

53 Denby Grange Colliery, Barnsley Area

SE 268 153

A number of small structures survive amid new development on the site of the former Denby Grange Colliery (centre), between Barnsley and Huddersfield. Started in 1894, two shafts (right centre) were sunk to depths of 171 yards (156m) and 164 yards (150m). Coal production was concentrated at these shafts, ventilation being provided by a 1200-yard (1097m) long drift. The mine was later linked to Caphouse Colliery (52). Set on the western side of the Barnsley area, the coal at Denby Grange was reached at a shallow depth. The coal seams in this area dip away from west to east and as a result mines in the eastern part of the coalfield are much deeper. South Kirby, near Pontefract, for example, reached a depth of almost half a mile. Here, however, the coal measures outcrop to form the boundary of the coalfield, beyond which millstone grit, the Pennines, wool, and cotton come into their own.

54 Emley Moor Colliery, Barnsley Area

SE 234 131

Situated near Huddersfield, Emley Moor mine stands in the northern part of the Barnsley coalfield in an area where there are a series of outcrops. As a consequence Emley Moor, in common with other neighbouring mines, was worked predominantly by drift. A shaft, 128 yards (117m) deep, gave access for men and materials while two surface drifts, one on another site, allowed coal to go directly to a coal preparation plant at Skelmanthorpe. Without coal handling or preparation equipment the mine site, as can be seen from the photograph, is very compact.

The entrance road from Chapel Lane (left) passes the car park, the cycle shed, the canteen, medical centre, and pit head baths (left centre). Beyond these (centre) are two buildings which contain various offices and the lamproom. Here the yard widens out, and the principal structures in the yard are the three contiguous buildings which housed the workshops, the large stores building and the compressor house, standing at the head of the hairpin-shaped road which enclosed the stockyard (centre).

55 Grimethorpe Colliery, Barnsley Area

SE 405 085

The name of Grimethorpe was synonymous with the Yorkshire coal industry. The last of Barnsley's mines to close, the photograph shows it as a massive industrial complex. Five substantial elements fill the landscape. In the foreground (right centre) is the site of Carlton Main Brickworks and above it the colliery itself with its No 3 Shaft winding gear (right) and the skip-winding facilities at the No 2 Shaft (right centre) standing in front of the power station with its two tall chimneys. To the left of the power station is a substation and beyond it the unmistakeable shape of a cooling tower (upper right centre). To the left of the reservoirs (centre) are tanks, ovens, chimneys and bunkers of the coalite plant. Finally the modern cladding of the coal preparation structures is visible across the rail tracks (upper left). The mine itself was sunk in 1894 and eventually mined to a depth of almost 1,000 yards (914m). In 1963 it was amalgamated with the nearby Houghton Main Colliery (56), and the production of both mines was brought to the surface at Grimethorpe.

58 Royston Drift Mine, Barnsley Area

SE 375 122

58.1 Royston Drift was established in 1976 on the site of the earlier mine, dating from the 1870s. The coal was worked via an intake drift and return drift to the east of the B6428 on the outskirts of the large South Yorkshire mining village of Royston. The two drifts, 434 yards (397m) and 422 yards (386m) in length, dipped at a gradient of 1:4 to the Sharlston Seam. These, like the concrete winding tower (upper centre) of the earlier shaft, are now disused. The twin ventilating 'chimneys' along with the forked chambers leading to the shaft from the fanhouse (upper centre) are typical of the ventilation arrangements installed by the National Coal Board at many mines.

58.2. Although the drift has closed, the coke, tar, and chemical works of the Monckton Coke and Chemical Company continue to function. The disused, low-level mine buildings stand above the road, but the dominating structures are within the working complex, particularly the tall black brick chimney (upper centre) and the wooden cooling tower (right). The constantly burning flame atop the waste gas chimney in the foreground (lower right) can be seen throughout the former mining village.

59 Woolley Colliery, Barnsley Area

SE 310 110

Massive modernisation of the processing plant in the 1980s made Woolley a prominent landmark fom the nearby M1 motorway. From above the circular thickeners, the tall loading bunkers and the square slurry ponds are clear (upper right centre and upper right). Less easy to distinguish but still visible are the baths, canteen complex (left), and the four bays of the stores standing at its side. The mine dated from 1869 when the outcropping Barnsley Seam was worked. The addition of shafts to lower seams in 1912 and 1942 expanded production at the site. A ventilation shaft on a separate site at Wheatley Wood was sunk from 1959 to 1961. The producing shafts at Woolley were located in the area in front of the pithead baths.

60 Askern Colliery, Doncaster Area

SE 561 135

The Doncaster Coalfield was opened out in the years from the turn of the century to the beginning of the First World War. It benefited from considerable investment from the National Coal Board. Askern, sunk between 1911 and 1916, working to a depth of 566 yards (517m), was one of the mines which was fully modernised both underground and on the surface. In common with nearly all the Doncaster mines, a new coal preparation plant (centre) was developed by the National Coal Board. This was followed by the introduction throughout the area of skip-winding facilities at the mouth of the producing shaft (left centre). Some of the details of the pre-modernisation plan of the colliery are still in evidence, for example, the railway lay-out near the coal preparation plant (right centre). As at collieries in other areas, the ancillary buildings are of a different vintage to the production and preparation plant, the offices and workshops near the pit yard entrance (left centre) being two examples. The pit head baths stand across the road from the entrance (left centre).

61 Bentley Colliery, Doncaster Area

SE 570 074

The development of the mines in the area around Doncaster was based on the highly lucrative Barnsley Seam. Rich and thick, between 6 and 14 feet (1.8 - 4.3m) high, the seam was the basis of the industry here for over 50 years. Bentley Mine was developed from 1905 to 1909 and found the 'Barnsley' at 624 yards (570m) below surface. In 1940 a further seam, the Dunsil Seam, was opened out and worked from the main level. A mixture of surface installations continued in use until the 1990s, skip-winding was introduced and some new preparation equipment was installed. The typical design of the tall loading bunker (centre) and the same functional design of the two winding houses (lower left centre) are illustrated clearly here. Many of the principal colliery buildings which developed at the time of, and in the years immediately after, the sinking were constructed of reinforced concrete using the Hennebique system. The engineer was L G Mouchel, Hennebique's agent in this country.

62 Brodsworth Colliery, Doncaster Area

SE 525 068

Pithead baths, offices, stores, and workshops buildings survive on the otherwise deserted site of Brodsworth Colliery. The colliery had previously worked via three shafts, two of which were sunk in 1905/07, the third in 1923. Here the Barnsley Seam was reached at a depth of 595 yards (544m), but later developments saw the shafts deepened to 842 yards (770m) to reach the Parkgate and Thorncliffe Seams. Now only the large baths building (lower right centre), the general offices (lower right), the stores (lower right) and the former joiners' and fitters' workshop (centre) testify to the site of the mine. A rectangular slurry pond is at the centre of the picture. An impressive, formally-arranged housing estate stands beside the pit site. Between the two lay the course of a Roman road which can be seen leading through the trees in the upper extreme left of the photograph.

65 Hatfield Colliery, Doncaster Area

SE 653 112

Hatfield Colliery was sunk in the same years as the Askern mine (60), in 1911-16. Here two winding engines and the power house (centre) have been retained among the modernised plant introduced by the National Coal Board. The two engine houses which probably date fom the sinking of the mine contrast sharply with the pair seen at neighbouring Bentley Colliery (61). The spider-like preparation plant (centre and left centre) leads to the loading bunker adjacent to the railway (centre). A set of waiting wagons is being filled. Rail distribution has been highly important to the Doncaster mines. At the beginning of the 1980s almost 90% of the area's coal was being handled by rail, the chief market for it being the power stations of South Yorkshire which took three-quarters of the coal mined in the Doncaster Area.

66 Hickleton Colliery, Doncaster Area

SE 465 053

Hickleton was for many years the only mine in Doncaster Area with nineteenth-century origins, having been sunk *c*1892 to the Barnsley Seam. By 1955 this seam was worked out and production was concentrated on others. Now production has stopped and much of the surface has been cleared. The shafts were still in operation, however, at the time of the photograph. The heapstead arrangement nearest to the camera probably dates from 1921 when a new shaft was sunk to a depth of 789 yards (721m) to reach the Parkgate Seam. The working out of the productive Barnsley Seam made Hickleton the smallest producing mine in the Doncaster Area.

67 Markham Main Colliery, Doncaster Area

SE 617 046

This photograph shows Markham Main shortly before its closure in late 1992. The mine was planned before the First World War but sinking was not completed until 1924 when the Barnsley Seam was reached at a depth of 730 yards (667m). The delay was due partly to world events and partly to the difficulties encountered in sinking through the highly porous strata, this being a problem encountered in the sinking of most of the Doncaster mines. The difficulties lay in the great depths which mines had to reach in order to find the coal, and the water-bearing rock which lay above it. Most of the mines had to adopt cast-iron tubbing blocks within the shafts to prevent water seeping in. These problems explain the comparatively late development of this part of the coalfield. The original planning of the mine is represented in the ordered rows of single-storey workshops (lower centre) which appear to be contemporary with the opening of the pit. The modern installations around the pit head (centre) are to the same design as those seen at a number of other mines in the area. Particularly prominent on this picture is the pithead baths building (left).

68 Rossington Colliery, Doncaster Area

SK 603 984

Two strikingly tall headstock towers (centre) form the focus of this complex where the survival of a number of buildings from the sinking of the mine in 1912-15 give it the appearance of a site in the older North East Area rather than in the highly mechanised Doncaster coalfield. Rossington is the deepest and the most southerly of the Doncaster mines and was, in the 1970s, the subject of a scheme for direct feeding of coal to a local power station using a new, rapid-loading bunker and a merry-go-round rail system. A feature rare to the modern mine site but still seen at Rossington is the tall chimney (centre). Also visible here is the forked ventilation chambers of the fan house (centre).

Map 6: Midlands Coalfields Areas

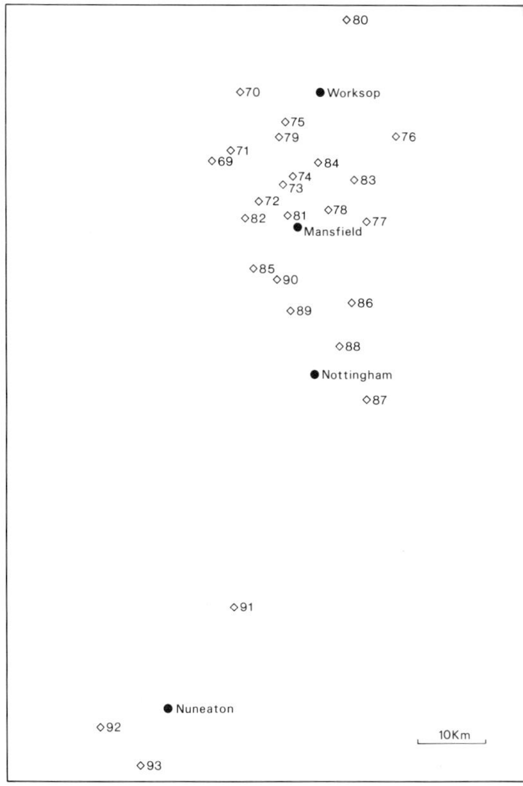

Hucknall colliery, South Nottinghamshire Area, 1955 (English Heritage)

69 Arkwright Colliery, North Derbyshire Area

SK 428 702

Looking north east across the almost deserted site of Arkwright Colliery, west of Chesterfield, the only surviving structure to be seen here are the pit head baths dating from 1954 and extended in 1976 (left centre). Arkwright was originally a small drift mine started in 1938 but was connected five years later to nearby Calow Colliery. Closure of a number of neighbouring mines in the late 1960s led to a concentration of activity in this part of the coalfield at Arkwright and the driving of a new drift in 1973-4. Only the mechanical and electrical engineers' workshops and part of the coal preparation plant were retained in the modernisation of the surface. Most of the surface structures were of a functional design, being brick-built or steel-framed and clad with corrugated sheeting. The building pad of the electrical and mechanical workshops is still visible (right centre).

70 High Moor Colliery, North Derbyshire Area

SK 468 800

The site of a 1950s drift mine opened by the National Coal Board, High Moor lies between Worksop and Sheffield. The buildings which still stand are of the most basic design, being predominantly brick with corrugated roof sheets. Production was via two drifts on the main site (centre) and a third in an adjacent area (upper right). To the north east (right) of the main access road can be seen the pithead baths and the pitched roofs of the offices and the canteen (right centre). The most prominent feature from the air is the row of four rectangular settling ponds (left), the two nearer the camera being for mine water, the others for surface water.

71 Markham Colliery, North Derbyshire Area

SK 455 718

By the end of the 1970s Markham near Bolsover was one of the largest and most modern mining complexes in the country. Sunk from 1882 to 1887, four shafts were at work within the pit yard. A massive £10 million reorganisation in 1976-7 brought skip winding, further coal preparation installations, and increased output from the underground link to Ireland Colliery. Despite these changes the surface retains some vestiges of its past. The oldest structure on site is the No 3 Shaft headstock. Of steel construction and brick cladding, it was constructed in 1891 (right). Nearby the No 2 Shaft headstock, erected in 1923, is of a standard steel lattice-work design. Surrounding these are a number of workshops dating from 1902 of brick construction with slate roofs and arched headed windows. A large compressor house dating from the 1920s stands at the rear of the No 1 headstock (left). The pit head baths in the foreground date from 1938 (centre).

The picture gives an opportunity to look at a mid-1970s coal preparation plant which, viewed from left to right, incorporates a dirt bunker, main coal preparation area with two circular thickening tanks in front, three cylindrical loading bunkers, the main washery (right centre), and a Bradford breaker at the rear of the site. There are numerous other components which make up the treatment plant which, via a merry-go-round rail system, fed directly to local power stations. Markham, together with the coalite plant just visible at the top of the photograph and Bolsover colliery beyond it, formed a major component of the national industry.

72 Pleasley Colliery, North Derbyshire Area

SK 496 645

At the disused Pleasley site, in contrast to many others, the coal production facilities survive while everything else has been swept away. The mine was sunk in 1873 to a depth of 698 metres and was one of the first mines in the country to use electricity underground. The mine worked the Deep Hard and the First Piper Seams. Retained on the former colliery surface (right) are the original engine house and the headstock and pit head structure of the No 1 Shaft dating from 1904 and the headstock and engine house of the No 2 Shaft dating from 1898. A substantial brick chimney which served the colliery's Lancashire boilers installed in the 1920s still towers above the buildings. Until its closure in 1983 the pit yard also contained workshops which had been erected in the 1890s.

73 Shirebrook Colliery, North Derbyshire Area

SK 530 670

This is one of a number of North Derbyshire collieries sunk in the 1890s and substantially modernised by the National Coal Board. Shirebrook lies north of Mansfield and was sunk in 1896, although no buildings now exist from that date. The oldest building on the site is the workshop block (centre) erected in 1920, a long building of stone construction with twelve roof bays. The stores (right centre) were added in 1922 and the pit head baths, lamproom, and canteen in 1932. Modernisation of the production area took place in the 1960s. The No 1 headstock, on the left as viewed here, winding engine and house, and heapstead were all erected in 1961, while those at the No 2 Shaft were constructed in 1966. A concrete and asbestos coal preparation plant was erected in 1963.

74 Warsop Main Colliery, North Derbyshire Area

SK 546 680

Sunk in 1893-5 Warsop Main Colliery was substantially reorganised in the 1950s. Only two buildings survived the modernisation scheme. The pit head baths and the medical centre, both brick-built, were erected in 1934. New headframes, winding engines, winding engine houses, ventilation houses, and offices were erected in 1958. The mine was based around two shafts, the caps of which can be seen (right centre) in front of the surviving workshops. The No 1 Downcast Shaft is nearer to the camera, the No 2 Upcast Shaft beyond. The workshops complex dating from 1965 contained a large engineering shop at its centre and surrounding it the former loco garage, mechanical store, electrical store, sub-station, electrical workshops, and joiners' shop. The second remnant is the office block which contained the time office and lamproom (right centre). The site of the H-shaped baths building can be picked out beyond the offices.

75 Whitwell Colliery, North Derbyshire Area

SK 535 758

Sunk in 1890, the oldest structure on the site was the stores building erected in 1895. The offices were housed in a building which had been the surface stables, also built before 1900. The introduction of electric winders and brick and concrete headframes and winding houses significantly altered the appearance of the colliery in the late 1950s. The lamproom, medical centre, coal preparation plant, and screens were also added in the 1950s. The canteen and the workshops were constructed in the 1960s. The pithead baths dated from 1935 while the ventilation house and the steam-raising plant were built in the 1940s.

Whitwell, between Worksop and Chesterfield, is yet another example of a site where the coal production and preparation equipment has been swept away but a number of reusable buildings remain. Perhaps unusually the pit head baths which stood on the east side of Southfield Lane (which runs across the centre of the picture) have been removed while a number of other buildings have been retained. The group which stands to the left of the former entrance to the mine (centre) comprise the former stores, oil store, heavy machinery store, sawmill and joiners' shop, garage, and rescue station. Those to the right of the entrance were the workshops and lamproom.

76 Bevercotes Colliery, North Nottinghamshire Area

SK 695 739

Bevercotes mine, near Retford, was sunk from 1953 to 1958 to a depth of 877m, working the Parkgate and Top Hard Seams. In its woodland setting the site is unusual but in its layout the colliery is very commonplace. Viewed from the north west the two heapsteads, No 1 standing to the right of No 2, are at the centre of the photograph. Coal is fed directly from the shaft mouth to the coal preparation plant (left) and then on to its various bunkers. Here, before reaching the tall clean-coal outloading bunker (centre) the coal passes through the ash and moisture determination house adjacent to the main preparation plant.

77 Bilsthorpe Colliery, North Nottinghamshire Area

SK 653 615

This is one of a number of mines in the North Nottinghamshire area which opened out in the 1920s. Sunk from 1925 to 1928 and working to a depth of 724m the plant was originally steam-powered. In common with all the area's mines the coal producing and preparation systems were modernised by the National Coal Board, in this case in the late 1970s. Electric winders were installed in 1978 and 1979 and a new coal preparation plant was opened in 1979-80. However, a number of buildings not directly linked with the extraction and processing of the coal survived the modernisation process. The lamp cabin, identifiable by the four openings in its gable end, at right centre, the range of joiners, blacksmiths, and electrical engineers' shops across the site road from the lamp cabin, all pre-date the reconstruction of the site. The No 2 Upcast Shaft is at the exact centre of the photograph with the No 1 Downcast Shaft directly south of it. The new coal preparation plant dominates the foreground. The long gantry which passes from left to right terminates at a dirt bunker. The rapid loader nearest the camera was constructed in 1976 to feed permanently coupled merry-go-round trains taking coal to the Trentside power stations.

78 Clipstone Colliery, North Nottinghamshire Area

SK 596 632

Approaching from the north, the camera gives a clear view of the surviving early buildings on this colliery site dating from 1920-2. As on many sites the buildings which have survived restructuring schemes are the ancillary ones. Here, at the entrance from Mansfield Road (centre) stand the offices, garage, control room, and the long single-storey building which contained the cable shop, the electricians' workshops, and the engineer's office. However it is the striking headstocks, some of the tallest in the country at 65m high, which dominate the site. The mine, which has two of the deepest shafts in the country, approximately 920m deep, stands cheek by jowl with the housing of New Clipstone which covers the lower half of the photograph.

79 Creswell Colliery, North Nottinghamshire Area

SK 524 736

Creswell Colliery was sunk in 1896. The aerial view captures the compact nature of the site. Although modernised in line with all the North Nottinghamshire mines a number of earlier buildings survive. However, more importantly this picture reflects the strong influence which the coal industry had on the development of settlements. At the right of the picture is the model village of New Creswell set out from the opening of the mine and completed at the turn of the century. The housing was set about an elongated octagonal space in an inner and an outer circuit. The inner space contained a formalised network of paths, a bandstand and a tree-lined avenue at the head of which stood the Workmen's Institute (right). In front of the Institute lie the bowling green and cricket pitch. Allotment gardens filled the area south west of the cricket pitch (right centre).

80 Harworth Colliery, North Nottinghamshire Area

SK 625 903

On the northern border of Nottinghamshire, some of the underground workings of Harworth Colliery lay beneath Yorkshire. Sunk in 1919-24 and mining to a depth of 894m, the site was considerably modernised in the 1980s. The restructuring of the surface created not only the new coal preparation plant with its typical tentacles (lower centre) and a rapid loading bunker (right) but also a new heapstead building for the No 1 Shaft (upper left) which towers over the adjacent earlier, possibly original, heapstead at the mouth of No 2 Shaft which can be seen with its typical forked arrangement of the fanhouse to the left of the centre of the photograph.

81 Sherwood/Mansfield Collieries, North Nottinghamshire Area

SK 537 624

One of Mansfield's two collieries, Sherwood opened in 1902-3, followed soon after by Mansfield Colliery in 1904. The last working steam winder, a double cylinder horizontal Fraser & Chalmers engine dating from the sinking of the mine, survived here until *c 1983*. This view of the mine is from the south. The substantial brick-built Edwardian office building guards the road entrance from the south. While the winding arrangements have been electrified many of the earlier surface buildings have been retained, typically the extensive workshop block at the centre of the photograph. The coal preparation plant was built in the 1960s and is, therefore, of a different generation to those at other Nottinghamshire mines. As a result there is not the proliferation of covered gantry conveyors snaking around and about the site as seen at many other collieries.

82 Silverhill Colliery, North Nottinghamshire Area

SK 475 620

At the time of the photograph Silverhill, west of Mansfield Woodhouse, was North Nottinghamshire's oldest working mine having started in 1875. Modernised facilities include a skip-winding system at the No 1 Shaft (centre) and a rapid-loading bunker (centre). The railway system was shared by nearby Teversal Colliery, the deserted site of which is starting to come into view in the lower extreme left. The 1918 Ordnance Survey map of the area shows banks of coke ovens on either side of the railway line. Silverhill Farm stands surrounded by trees in the fields directly behind the mine. The agricultural landscape in the background contrasts sharply with the rugged relief of coal tips, slurry lagoons and former clay pits in the foreground.

83 Thoresby Colliery, North Nottinghamshire Area

SK 635 675

By the time of the major investment programme in the early 1980s, Thoresby was one of the largest producing pits in the country with an output in the region of 1,800,000 tons worked from three seams to a depth of almost 700m. Large 500-ton and 1000-ton capacity vertical bunkers had been built underground, while on the surface new electric winders were installed at both shafts and a new rapid-loading bunker (left) was completed. Despite this a number of buildings dating from the opening of the mine in 1925-8 continued to be used. The long workshop building at the rear of the site (lower centre) contained the blacksmith's shop, the fitting shop, welding shop and the electricians' shop. The low single-storey building parallel to the rapid loading bunker housed the time and wages office, as well as the rescue room and the lamproom. Perhaps most surprising is that, despite the huge output of the mine, the structures at the shaft mouths remained largely unaltered.

84 Welbeck Colliery, North Nottinghamshire Area

SK 582 700

Sunk in 1912-15, virtually no structures remain on site from the First World War era, although the arrangement of the heapstead at No 1 Shaft (centre) is similar to the original layout. A long colliery workshops building survived for many years until a new building on the same site (right centre) was erected to provide modern facilities. By the early 1980s a stockpiling system had been introduced to allow the 1.1 million tons of production to be handled by the 1960s coal preparation plant (left centre). Although no signs of their existence can now be seen, the site also contained two brick kilns which stood near the site of the tall rapid-loading bunker on the left of the photograph.

35 Bentinck Colliery, South Nottinghamshire Area

SK 488 550

A major Nottinghamshire mining complex, near Kirkby in Ashfield, this colliery consistently produced over a million tons of coal from 1969 onwards. It was opened with the sinking of three shafts by the New Hucknall Colliery Co Ltd in 1895, but the shaft winding of coal stopped in 1968 when a new drift was driven into the Blackshale Seam. The drift mine entrance is on the left of the complex (left centre) from where coal is fed via conveyor belt to the screen house and to the massive coal preparation plant at the centre of the picture and of the complex. A high-level gantry leads to the tall rapid loading bunker. To the right of the bunker stands the surviving No 1 headstock and 1895 winding engine house which held a 1915 John Warner & Sons double horizontal steam engine with an 18 foot (5.5m) diameter drum.

86 Calverton Colliery, South Nottinghamshire Area

SK 603 500

86.1 The colliery itself appears dwarfed by the adjacent coal preparation plant. The complex, seven miles north of Nottingham, straddles Oxton Road with the mine (right) to the south and the preparation plant to the north (centre and left). The lines of the rail complex which formerly served the mine and the plant can still be detected at upper centre and around the main coal preparation building (left centre)

86.2 The headstocks of the No 1 and No 2 Shafts can be clearly seen at the centre of the photograph. The No 2 Shaft with its enclosed conveyor feeding a vertical bunker is the nearer to camera. The fan house nestles between the trees (lower centre). The winding houses for the shafts stand directly adjacent to the headstocks. Various stores and compounds spread eastwards toward to the top of the picture. The pit head baths, distinguishable by the square tower, stand on the right near the entrance to the pit yard from Hollinwood Lane (right centre). The uncluttered and functional complex of brick and concrete structures reflects its mid twentieth-century origins.

87 Cotgrave Colliery, South Nottinghamshire Area

SK 651 364

87.1. Only a handful of miles south east of the City of Nottingham, Cotgrave is a modern mine opened by HRH Princess Margaret. The long-range image is of industry in an agricultural setting. The camera angle has flattened the coal and waste heaps but the towers of the No 1 Upcast Shaft (closer to camera) and the No 2 Downcast Shaft rise out of the landscape.

87.2. Close up the details of the colliery are more prosaic. The coal is drawn up the No 1 Shaft by the winding gear within the tower (upper centre) and passes via a gantry to the coal preparation plant and washer (left) and to the tall rapid-loading bunker. The colliery workshops stand beside the No 2 Shaft (centre) and the pit head baths and canteen are on the right of the picture.

88 Gedling Colliery, South Nottinghamshire Area

SK 612 438

The colliery site, north east of Nottingham, is viewed from the west during demolition. Through the centre of the site ran the railway, discernible by the thin line of trees and vegetation running from left to right across the picture. Below this line the rubble of the baths, canteen, and lamproom buildings lie in front of the surviving offices and workshops (lower centre). The bridges from here to the main producing area of the mine stand at the centre of the photograph. A portion of the screens and washery building still stands (right centre) but the matching pair of winding engine houses which worked the two shafts have already been demolished, leaving only two concrete structures intact.

89 Hucknall Colliery, South Nottinghamshire Area

SK 540 490

Hucknall Colliery was sunk in the 1860s. A major surface reconstruction took place from 1957 to 1968, leaving only two buildings, the offices and the workshops, from the earlier period. The reconstruction itself produced a series of surprisingly striking concrete headstock structures, two of which (Nos 3 and 4) were connected and worked from the same electric winding engine house. The site is now cleared but the footprints of a number of buildings are visible from the air. The main producing and preparation area was at the centre of the photograph. The circular shape of the former water tank is evident to the left of this, and the outlines of the compounds and workshops can be seen nearby. Four sets of parallel lines at the centre indicate the rail complex which ran through the coal handling and preparation area.

90 Newstead and Annesley Collieries, South Nottinghamshire Area

SK 520 532 and SK 522 533

90.1 These two adjacent collieries less than a quarter of a mile apart date from the 1860s: the buildings and headstocks of Annesley (centre) represent continuing life at the time of this photograph, while the cleared site of Newstead (upper centre) records a different fate. Annesley Colliery was sunk from 1869 to 1875. The office near the car park and stone-built slate-roofed loco shed attached to the workshop and stores building, directly right of the car park, are Victorian survivals. The light colour of the tower of the No 2 Shaft catches the light and the eye at the centre of the photograph. Parallel lies the No 1 Shaft and to the right of each shaft is a functional, rectangular, electric winding-engine house. The V-shaped design of the baths, canteen, and medical centre forms a striking feature at lower left. Colliery housing which formed the village of Newstead at the sinking of the mines stood in a series of rows in the now cleared field in the upper right of the picture.

90.2. At the Newstead site only a handful of structures have been maintained, principally the baths building (centre) dating from 1930. A row of colliery houses stand across the now disused railway route, on which some lengths of track still remain.

91 Bagworth Colliery, South Midlands Area

SK 458 068

To the west of Leicester lie the mining complexes of Bagworth, Thornton, and Nailstone. The site of Bagworth Colliery is small scale. Two headstocks, unusually angled in relation to each other, stand at the centre of the site. The survival of a tall chimney (centre) on a colliery site is now rare. The rest of the surface is dedicated to the winding and ventilation facilities for the shafts, a small number of workshops, and the usual office and bath facilities.

92 Daw Mill Colliery, South Midlands Area

SP 260 900

92.1. Daw Mill mine at Arley between Birmingham and Nuneaton is a large modern installation where two shafts, the headstock of one of which can be seen at lower right, feed coal to the extensive coal preparation facilities. A prominent surface feature here is the large circular 'filter bed' used as part of the treatment process (centre).

92.2. A general view of the colliery shows in more detail the shaft top arrangements (right and right centre) and the large coal preparation plant. The view also gives an indication of the amount of stock, stores, and materials used in modern mining, which cover large areas of the surface of the yard.

93 Keresley Colliery, South Midlands Area

SP 320 844

Keresley Colliery, near the M6 motorway on the north western edge of Coventry, illustrates many of the facets of the twentieth-century development of the colliery surface. The photograph shows older headstocks (right centre and right) served by modern winding-engine houses (centre and right centre). Workshops, stores, and offices which pre-date the nationalisation of the industry (left centre and left) and, in the foreground, the washery, covered conveyors, and bunkers typical of the coal preparation plants installed in the 1960s and 1970s. A further commonplace but as yet unremarked element of the surface facilities at this and every other mine is the pit car park (upper centre), a symbol of the changed nature of the industry. For over a century mining was at the heart of the development of local communities, villages, and towns throughout the nation's coalfields. The considerable decline in the number of mines, industrial diversification in the coal areas, and extensive car ownership has altered that relationship. Ominously, however, the car park is empty.

Glossary of mining terms

adit: a tunnel rising at 1:100 or 1:200 for free drainage and / or to assist in the transportation of loaded waggons.

coke oven: coking, the carbonisation of coal, was originally performed in open heaps but by the nineteenth century brick-built beehive ovens became common and were often constructed in long banks. These were replaced during the twentieth century by by-product recovery ovens where the gas and tar driven off during the coking process were systematically recovered.

compressor house: a building housing a steam engine or other plant to produce compressed air for the operation of pneumatic machinery underground, introduced from the mid nineteenth century. During the early twentieth century, compressor houses were gradually replaced by generator houses as collieries became electrified.

corves: wicker containers used for transporting coal manually before animal traction or mechanisation were introduced.

downcast: see shaft

drift: a regular straight mine entrance, driven on a constant downwards slope, predominantly a twentieth-century technique.

electricity substation: see generator house.

engine house: a building housing an engine, usually with an attendant boiler house and chimney, for pumping water or winding from the pit, also known, in the latter use, as a winding or winder house. Steam engines, first the atmospheric (Newcomen-type) engine, introduced from the early eighteenth century, and then beam engines (eg Boulton and Watt, and Cornish-type) from the 1770s, remained in use well into the twentieth century. An important nineteenth-century variant, dominant in the north east, was the vertical winding engine house in which the winding drum was located over the cylinder of the engine, with no beam. Elsewhere in the country the horizontal steam engine was the norm, consisting of two horizontal cylinders attached by a crank to the winding drum and housed in a horizontal engine house. Gas, oil, and electrically powered engines appeared in the first decades of this century.

fan house: a building housing a mechanical fan, usually steam-driven until the introduction of electricity, to ventilate the underground workings and prevent the build-up of potentially explosive methane gas. By the mid nineteenth century ventilation was efficiently managed using ventilation furnaces at the base of the shaft, and a ventilation chimney over, or close to, the upcast shaft, to provide a greater draught. The risk of the furnace itself igniting a fire or explosion led to the introduction of mechanical ventilation.

generator house: a building where electricity was generated to power machinery, both above and below ground, notably winding engines, pumping engines, ventilation fans, and coal cutting equipment. See **compressor house.**

headgear or headstock: the pit top structure housing the winding gear. Early wooden examples were gradually replaced by iron, steel (girder and lattice construction), and ferro-concrete frames. Twentieth-century headgear arrangements include electric friction winders mounted directly over the shaft (where two cages are attached to a single endless rope).

heapstead: a group of structures or features around the mouth of the winding shaft, including a series of rail lines to convey coal tubs from the cage to the screens. The colliery headgear usually stands on the heapstead.

horse gin: primitive collieries often harnessed horse power for shaft winding and pumping. A horse or pair of horses were attached by a wooden pole to a horizontal drum and, as they walked around the rotating drum, a rope passed over a pulley beside the shaft and down into the mine. Other variants included the cog and rung gin.

incline: an embankment or cutting surfaced at a uniform gradient, up or down which materials were hauled by rope; normally part of a railway system.

lamproom: a building, also sometimes known as a lamp cabin, where the miners' lamps were stored and collected before going down the pit. Sir Humphry Davy's new safety lamp was introduced in 1816 and gained widespread use through the nineteenth century. Electric lamps did not become widespread until after the First World War. With the construction of pithead baths from the late 1920s, new lamprooms, where electric lamps could be cleaned, stored, and recharged after use, were commonly built as part of the baths complex.

outcropping: an irregular area of primitive open workings along the outcrop of a coal seam. Outcropping can occur over extensive areas and appears physically as a series of 'humps and bumps' in the ground.

pithead baths: a building containing the colliery showers and often the first-aid and mines rescue facilities. Pithead baths come in many shapes and sizes, most are characterised by the striking use of red brick and the prominent display of chimney and water tank. They are mainly a twentieth-century feature, although there

Allerton Bywater, North Yorkshire: the washery, one of the earliest electric powered plants in Britain, installed in the period before the First World War

is reference to the provision of warm baths and showers as early as 1855 at Cramlington Colliery in the north east. One of the earliest twentieth-century pithead baths was at Gibfield Colliery at Atherton, Greater Manchester, built in 1913, and hundreds more followed the provisions of the Mining Industry Act of 1926. Many were commissioned by the Miners' Welfare Committee and were designed by leading architects of the Modern Movement.

powder house: a storehouse for gunpowder or other explosives, typically thick- or double-walled, windowless, and located at some distance from the other colliery buildings.

preparation area or building: location of coal washing and screening processes. See **washery** and **screening.**

screening or screens: a method of grading coal, removing small pieces and waste. As introduced in the mid eighteenth century, the screen consisted of a series of fixed longitudinal bars of wood (later iron) onto which the coal was tipped. The small coal fell through the screen and was often used in the colliery boilers for steam raising. By the 1840s, complex wooden screening structures were attached to the heapsteads of larger collieries and towards the end of the century the process became increasingly mechanised, with picking belts (conveyors on which the coal was picked over manually) and jigging screens. The screens now means a building which contains the various forms of equipment for coal sizing and grading.

shaft: a vertical or sub-vertical opening from the surface, often lined with wood, stone, brick, or concrete, and used for access, haulage, ventilation, and/or pumping. The use of forced ventilation required the presence of two shafts. One, the downcast, was for air drawn down in to the mine. The other, the upcast, was for the air returning to the surface. These might be made by dividing a single shaft in two by a central partition. After the Hartley Colliery disaster of 1862, where 204 men and boys lost their lives, legislation was passed making two shafts compulsory.

shaft-mound: a shaft mouth surrounded by a ring of spoil and typically surviving as an earthwork with an inverted-bell profile (commonly referred to as a 'bell-pit'). Similar surface features may be the product of primitive, single shaft workings or complex underground galleries.

skip winding: adopted after nationalisation in 1947, skip winding was a more efficient alternative for raising coal to the use of tubs raised in the pit cage. Skip capacity was greater and loading and unloading was quicker. Despite this, its introduction was slow and in 1955 still only 5% of national output was being wound in skips. The massive skip winding towers of ferro-concrete constructed at new collieries had a dramatic visual impact.

spoil heap: conical or flat-topped tips of waste discarded from the mine or coal preparation plant became a characteristic feature of colliery landscapes. Waste was deposited by the use of tubways, aerial ropeways or conveyors.

upcast: see shaft.

ventilation chimney: a tall stack for the dispersal of foul air from the bottom of the mine. It normally implies the presence of a ventilation furnace positioned either at or below the surface. See **fan house**.

waggonway: a railed transport system, of any gauge and either gravity- or horse- drawn, rails were of wood or iron, and stone sleepers or blocks were used for anchorage. The linear earthwork on which the rails were seated is usually the only surviving feature.

washery: a building where coal is washed, sized, and graded before being sold, usually a substantial shed-like structure. Washing of coal was introduced in the 1880s but adopted slowly and only on a large scale since nationalisation in 1947. The coal washery has become increasingly important in recent years as many modern pits tend to extract entire seams including a great deal of dirt and small coal.

winder or winding house: see **engine house.**

workshops: a range of ancillary buildings commonly found on collieries, including the blacksmith's shop, fitting shop, joiners' shop, sawmill, and tub shop. These formed extensive brick-built complexes on twentieth-century sites.

Index of survey sites

The sites illustrated in the gazetteer are indicated by their reference number in the left hand column. Other sites included in the survey, but not included in the Gazetteer, are listed with an Ordnance Survey grid reference and accession number for the RCHME's National Library of Air Photographs (see p18 for details). For most sites there are multiple record photographs, taken from different aspects, and the single reference number given here is intended simply as an aid to locating other views in the Library accession register

Gazetteer number	Colliery name	NCB area	OS grid ref	RCHME acc no
1	Agecroft Colliery	West	SD 803 014	14371/8
32	Allerton Bywater Colliery	N Yorks	SE 424 277	14356/2
90.1	Annesley Colliery	S Notts	SK 520 532	14235/10
69	Arkwright Colliery	N Derbs	SK 428 702	14230/6
-	Ashby Colliery	S Midlands	SK 37 16	4746/45
14	Ashington Colliery	NE	NZ 262 880	14066/31
60	Askern Colliery	Doncaster	SE 561 135	14249/9
-	Babbington Colliery	S Notts	SK 533 438	14236/14
91	Bagworth Colliery	S Midlands	SK 458 068	4746/71
-	Barnburgh Colliery	S Yorks	SE 475 034	14242/8
50	Barrow (Barnsley Main) Colliery	Barnsley	SE 367 063	14247/6
15	Bates Colliery	NE	NZ 303 823	14344/3
-	Bearpark Colliery	NE	NZ 242 434	14066/47
85	Bentinck Colliery	S Notts	SK 488 550	14235/8
61	Bentley Colliery	Doncaster	SE 570 074	14244/4
76	Bevercotes Colliery	N Notts	SK 695 739	14229/9
2.1	Bickershaw Colliery	West	SJ 634 997	14372/1
77	Bilsthorpe Colliery	N Notts	SK 653 615	14234/1
-	Blidworth Colliery	N Notts	SK 595 565	14234/5
3	Bold Colliery	West	SJ 550 930	14367/12
16	Brenkley Colliery	NE	NZ 235 741	14066/20
62	Brodsworth Colliery	Doncaster	SE 525 068	14243/8
51	Bullcliffe Wood Colliery	Barnsley	SE 285 145	14251/7
86	Calverton Colliery	S Notts	SK 603 500	14236/6
52	Caphouse Colliery	Barnsley	SE 253 165	14250/14
-	Chatterley Whitfield Colliery	West	SJ 884 534	14375/1
78	Clipstone Colliery	N Notts	SK 596 632	14233/12
-	Coalville Colliery	S Midlands	SK 43 11	4746/49

Gazetteer number	Colliery name	NCB area	OS grid ref	RCHME acc no
40	Cortonwood Colliery	S Yorks	SE 408 014	14242/5
87	Cotgrave Colliery	S Notts	SK 651 364	14237/8
79	Creswell Colliery	N Notts	SK 524 736	14229/14
4	Cronton Colliery	West	SJ 475 894	14366/2
-	Darfield Main Colliery	Barnsley	SE 402 040	14246/15
92	Daw Mill Colliery	S Midlands	SP 260 900	14379/10
17	Dawdon Colliery	NE	NZ 435 480	14351/2
53	Denby Grange Colliery	Barnsley	SE 268 153	14251/1
41	Dinnington Colliery	S Yorks	SK 518 865	14239/2
-	Dodworth Colliery	Barnsley	SE 312 059	14247/11
18	Easington Colliery	NE	NZ 438 441	14352/1
19	Elemore Colliery	NE	NZ 355 457	14347/8
20	Ellington Colliery	NE	NZ 284 918	14342/7
-	Elsecar Colliery	S Yorks	SE 386 008	14246/7
54	Emley Moor Colliery	Barnsley	SE 234 131	14251/8
21	Eppleton Colliery	NE	NZ 363 483	14347/14
5	Florence Colliery	West	SJ 915 420	14377/3
63	Frickley Colliery	Doncaster	SE 445 117	14248/13
33	Fryston Colliery	N Yorks	SE 458 270	14356/12
88	Gedling Colliery	S Notts	SK 612 438	14237/4
-	Glass Houghton Colliery	N Yorks	SE 435 244	14357/5
6	Golborne Colliery	West	SJ 603 987	14369/14
64	Goldthorpe Colliery	Doncaster	SE 469 042	14243/2
55	Grimethorpe Colliery	Barnsley	SE 405 085	14248/8
7	Haig Colliery	West	NX 967 176	14381/3
-	Hapton Valley Colliery	West	SD 809 314	14382/13
80	Harworth Colliery	N Notts	SK 625 903	14245/11
65	Hatfield Colliery	Doncaster	SE 653 112	14249/2
28	Hawthorne Combined Mine	NE	NZ 385 45 3	14348/3
8	Hem Heath Colliery	West	SJ 885 415	14376/8
-	Herrington Colliery	NE	NZ 344 533	14346/9
66	Hickleton Colliery	Doncaster	SE 465 053	14242/13
70	High Moor Colliery	N Derbs	SK 468 800	14228/14
-	Holditch Colliery	West	SJ 839 477	14375/13
-	Horden Colliery	NE	NZ 444 418	14353/3
22	Houghton Colliery	NE	NZ 338 502	14346/11
56	Houghton Main Colliery	Barnsley	SE 420 062	14247/1
89	Hucknall Colliery	S Notts	SK 540 490	14246/1

Gazetteer number	Colliery name	NCB area	OS grid ref	RCHME acc no
23	Hylton Colliery	NE	NZ 365 584	14345/13
-	Ireland Colliery	N Derbs	SK 437 741	14229/15
34	Kellingley Colliery	N Yorks	SE 525 236	14354/11
93	Keresley Colliery	S Midlands	SP 320 844	14380/1
-	Kilnhurst Colliery	S Yorks	SK 460 969	14241/15
42	Kiveton Park Colliery	S Yorks	SK 493 827	14238/7
9	Lea Hall Colliery	West	SK 058 168	14378/9
35	Ledston Luck Colliery	N Yorks	SE 431 307	14355/9
-	Linby Colliery	S Notts	SK 536 505	14236/2
24	Lynemouth Colliery	NE	NZ 297 904	14342/14
43	Maltby Colliery	S Yorks	SK 552 925	14241/2
81	Mansfield Collliery	N Notts	SK 537 624	14233/3
44	Manton Colliery	S Yorks	SK 590 780	14228/5
49	Manvers Colliery	S Yorks	SE 446 010	14240/13
71	Markham Colliery	N Derbs	SK 455 718	14230/5
67	Markham Main Colliery	Doncaster	SE 617 046	14244/8
-	Marley Hill Colliery	NE	NZ 215 575	14066/35
-	Moorgreen Colliery	S Notts	SK 480 477	14236/12
25	Murton Colliery	NE	NZ 399 472	14348/10
-	Nailstone Colliery	S Midlands	SK 42 08	4745/1
90	Newstead Colliery	S Notts	SK 522 533	14235/12
57	North Gawber Colliery	Barnsley	SE 330 008	14247/14
-	Nostell Colliery	N Yorks	SE 400 170	14250/4
45	Orgreave Colliery	S Yorks	SK 426 870	14238/9
-	Park Mill Colliery	Barnsley	SE 258 118	14251/10
10	Parkside Colliery	West	SJ 600 947	14368/8
11	Parsonage Colliery	West	SD 653 009	14372/11
72	Pleasley Colliery	N Derbs	SK 496 645	14232/14
36	Prince of Wales Colliery	N Yorks	SE 452 229	14357/12
57	Redbrook Colliery	Barnsley	SE 330 008	14247/14
-	Renishaw Park Colliery	N Derbs	SK 438 774	14229/2
-	Rockingham Colliery	Barnsley	SE 350 001	14246/12
68	Rossington Colliery	Doncaster	SK 603 984	14244/15
58	Royston Drift Mine	Barnsley	SE 375 122	14358/13
-	Sacriston Colliery	NE	NZ 235 478	14066/38
26	Seaham Colliery	NE	NZ 413 496	14349/15
37	Selby Project/ Gascoigne Wood Drift	N Yorks	SE 523 318	14364/10

Gazetteer number	Colliery name	NCB area	OS grid ref	RCHME acc no
38.1	Selby Project/North Selby Mine	N Yorks	SE 640 440	14362/11
38.2	Selby Project/Riccall Mine	N Yorks	SE 640 360	14361/5
38	Selby Project/Shaft Mines	N Yorks	SE 523 318	14364/10
38.3	Selby Project/Stillingfleet Mine	N Yorks	SE 600 410	14363/1
38.4	Selby Project/Whitemoor Mine	N Yorks	SE 665 356	14360/1
38.5	Selby Project/Wistow Mine	N Yorks	SE 575 355	14363/11
39	Sharlston Colliery	N Yorks	SE 361 180	14250/9
81	Sherwood Colliery	N Notts	SK 537 624	14233/3
27	Shilbottle Colliery	NE	NU 210 080	14066/4
73	Shirebrook Colliery	N Derbs	SK 530 670	14232/8
-	Shireoaks Colliery	S Yorks	SK 560 810	14238/3
12	Silverdale Colliery	West	SJ 815 467	14374/14
82	Silverhill Colliery	N Notts	SK 475 620	14234/7
46	Silverwood Colliery	S Yorks	SK 478 941	14241/9
63	South Elmsall Colliery	Doncaster	SE 445 117	14248/13
28	South Hetton Colliery	NE	NZ 385 453	14348/3
-	South Kirby Colliery	Barnsley	SE 465 097	14249/11
-	Sutton Colliery	N Notts	SK 484 603	14234/14
-	Sutton Manor Colliery	West	SJ 517 908	14366/10
-	Thornton Colliery	S Midlands	SK 45 06	4746/53
83	Thoresby Colliery	N Notts	SK 635 675	14231/9
47	Thurcroft Colliery	S Yorks	SK 500 885	14239/10
48	Treeton Colliery	S Yorks	SK 435 878	14238/15
29	Vane Tempest Colliery	NE	NZ 423 503	14350/15
74	Warsop Main Colliery	N Derbs	SK 546 680	14232/2
49	Wath Colliery	S Yorks	SE 446 010	14240/13
23	Wearmouth Colliery	NE	NZ 393 580 -	
84	Welbeck Colliery	N Notts	SK 582 700	14231/4
13	West Cannock No 5 Colliery	West	SK 008 141	14378/14
30	Westoe Colliery	NE	NZ 375 670	14344/11
-	Westthorpe Colliery	N Derbs	SK 452 795	14228/15
-	Wheldale Colliery	N Yorks	SE 441 262	14356/6
-	Whittle Colliery	NE	NU 175 065	14066/9
75	Whitwell Colliery	N Derbs	SK 535 758	14229/3
-	Wolstanton Colliery	West	SJ 860 480	14375/1
31	Woodhorn Colliery	NE	NZ 290 885	14343/8
59	Woolley Colliery	Barnsley	SE 310 110	14252/2
-	Yorkshire Main Colliery	Doncaster	SK 543 992	14240/2